INTERMEDIATE ALGEBRA

A BASIC APPROACH

Second Edition

Magdala Emmanuel

VALENCIA COLLEGE

Linus
Publications, Inc.

Published by Linus Publications, Inc.
Ronkonkoma, NY 11779

ISBN 10: 1-60797-436-3

ISBN 13: 978-1-60797-436-9

Printed in the United States of America.

This book is printed on acid-free paper.

Print Number   5   4   3   2   1

A letter from the author

Dear Instructors,

I want to thank you for using my book. While you may prefer to use a book that has fancy features and graphics, you are using this one. I am very grateful, and your students will also be grateful to you for the cost savings opportunity that this book affords them. It is significantly cheaper than the previous ones used.

This book addresses basic skills that will prepare students for the next level algebra course. All the skills needed for this level are addressed (Including the use of graphing calculators). The approach to this book is straightforward: each topic is briefly explained, then followed by worked examples. Throughout the book, I strive to keep the methods consistent. For instance, I consistently use the point-slope form of the equation whenever I need to find an equation of the line. For factoring trinomials, where three methods can be used, I use the grouping method because that method was introduced earlier in the section. For simplifying complex fractions and solving rational equations, I consistently use one method, clearing the fractions. A few of you may be partial to synthetic division; however, I use long division because it works for all types of divisors. Nonetheless, you are welcome to use whichever methods that you are comfortable with and that are beneficial for your students. Most sections are not overloaded with information, so you should have enough time for practice at the end of each class period.

When creating the applications, I strived to incorporate things that are familiar to the students. The applications cover topics such as working in the tutoring center, paying off a school loan, dropping an object from the clock tower, majoring in mathematics, working together on a job, selling graphing calculators, charging admission to a benefit concert on campus, traveling to a Phi Theta Kappa convention, playing soccer, predicting the growth of the Supplemental Learning program, scoring algebra tests, and many more.

I hope you will like using the book. Thanks again for the support.

Sincerely yours, Magdala

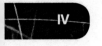

Table of Contents

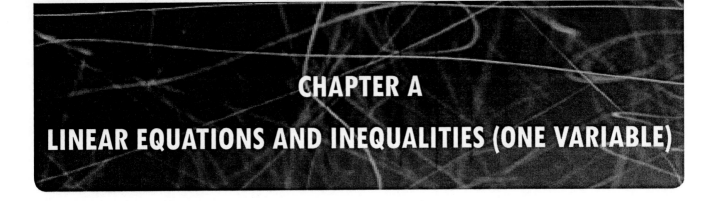

CHAPTER A

LINEAR EQUATIONS AND INEQUALITIES (ONE VARIABLE)

Section A.1 – Absolute Value Equations

❖ Solving absolute value equations

❖ *Solving absolute value equations*

Before solving absolute value equations, let us define the absolute value of a number. The absolute value of a number can be viewed as the distance of the number from zero or how far the number is from zero (not which direction). We will solve equations that have the absolute value of a variable expression.

For example, the absolute value equation $|x| = 3$ has for solution all real numbers which distance from 0 is 3 units. Both -3 and 3 are 3 units away from 0 on the number line.

Below are examples of absolute value equations:

$$|x| = 4$$

$$|x + 1| = 6$$

$$|2x + 1| = |x - 5|$$

Example 1

Solve the absolute value equation: $|x| = 4$.

The solutions are: -4 and 4. To check the solutions, replace them in the original equation.

Example 2

Solve the absolute value equation: $|x + 3| = 2$.

We set the expression inside the absolute value symbol equal to -2 and 2.

$x + 3 = 2$ $x + 3 = -2$

$x = 2 - 3$ (Subtract 3 from both sides.) $x = -2 - 3$ (Subtract 3 from both sides.)

$x = -1$ $x = -5$

The solutions are: -5 and -1.

To check the solutions, replace them in the original equation.

Example 3

Solve the absolute value equation: $|4x - 1| = 6$.

We set the expression inside the absolute value symbol equal to -6 and 6.

$4x - 1 = 6$ $4x - 1 = -6$

$4x = 6 + 1$ (Add 1 to both sides.) $4x = -6 + 1$ (Add 1 to both sides.)

$4x = 7$ $4x = -5$

$x = \dfrac{7}{4}$ (Divide by 4.) $x = \dfrac{-5}{4}$ (Divide by 4.)

The solutions are: $-\dfrac{5}{4}$ and $\dfrac{7}{4}$.

To check the solutions, replace them in the original equation.

Example 4

Solve the absolute value equation: $\left|\dfrac{x}{5}\right| = 10$.

We set the expression inside the absolute value symbol equal to -10 and 10.

$$\dfrac{x}{5} = 10 \qquad\qquad\qquad \dfrac{x}{5} = -10$$

Clear the fractions.

$$\dfrac{5(x)}{5} = 10(5) \text{ (Multiply both sides by 5.)} \qquad \dfrac{5(x)}{5} = -10(5) \text{ (Multiply both sides by 5.)}$$

$$x = 50 \qquad\qquad\qquad\qquad x = -50$$

The solutions are: -50 and 50.

To check the solutions, replace them in the original equation.

Example 5

Solve the absolute value equation: $\left|\dfrac{2x+7}{3}\right| = 7$.

We set the expression inside the absolute value symbol equal to -7 and 7.

$$\dfrac{2x+7}{3} = 7 \qquad\qquad\qquad \dfrac{2x+7}{3} = -7$$

Clear the fractions.

$$\dfrac{3(2x+7)}{3} = 7(3) \text{ (Multiply both sides by 3.)} \qquad \dfrac{3(2x+7)}{3} = -7(3) \qquad \text{(Multiply both sides by 3.)}$$

$$2x + 7 = 21 \qquad\qquad\qquad\qquad 2x + 7 = -21$$

$2x = 21 - 7$ (Subtract 7 from both sides.) $\qquad 2x = -21 - 7$ (Subtract 7 from both sides.)

$2x = 14 \qquad$ (Combine like terms.) $\qquad 2x = -28 \qquad$ (Combine like terms.)

$x = 7 \qquad$ (Divide both sides by 2.) $\qquad x = -14 \qquad$ (Divide both sides by 2.)

The solutions are: -14 and 7.

To check the solutions, replace them in the original equation.

Example 6

Solve the absolute value equation: $|x + 2| + 14 = 4$.

Rewrite the equation by subtracting 14 from both sides: $|x + 2| = 4 - 14 \rightarrow |x + 2| = -10$

The absolute value of a number is never negative. This equation has **no solutions**.

Example 7

Solve the absolute value equation: $|x + 1| = |6x - 4|$.

We set the expression inside the absolute value symbol, on the left side of the equation, equal to $(6x - 4)$ and to $-(6x - 4)$,

$x + 1 = 6x - 4$ $\qquad\qquad\qquad\qquad$ $x + 1 = -(6x - 4)$

$x + 1 = 6x - 4$ (No change) $\qquad\qquad$ $x + 1 = -6x + 4$ (Distribute the -1.)

$x - 6x + 1 = -4$ (Subtract $6x$ from both sides.) $\qquad$ $x + 6x + 1 = 4$ (Add $6x$ to both sides.)

$-5x = -4 - 1$ (Subtract 1 from both sides.) $\qquad$ $7x = 4 - 1$ (Subtract 1 from both sides.)

$-5x = -5$ (Combine like terms.) $\qquad\qquad$ $7x = 3$ (Combine like terms.)

$x = 1$ (Divide both sides by -5.) $\qquad\qquad$ $x = \dfrac{3}{7}$ (Divide both sides by 7.)

The solutions are: $\dfrac{3}{7}$ and **1**.

To check the solutions, replace them in the original equation.

Section A.1 - Absolute Value Equations

✎ Your turn...

Solve the absolute value equations.

1. $|x| = 8$

2. $|x + 7| = 5$

3. $|3x - 2| = 9$

4. $\left|\dfrac{x}{4}\right| = 12$

5. $\left|\dfrac{3x-7}{2}\right| = 5$

6. $|x + 1| + 10 = 2$

7. $|4x + 9| = |x - 5|$

8. $|x - 2| - 5 = 11$

For more practice, access the online homework. See your syllabus for details.

Section A.2 – Compound Inequalities

- ❖ Solving compound inequalities (AND)
- ❖ Solving compound inequalities (OR)

When two inequalities are separated by the words: **or, and,** they are called compound inequalities.

Below are examples of compound inequalities:

$$2x + 3 > 7 \text{ and } x < 5$$

$$x + 3 \le 17 \quad \text{or} \quad 2x < \frac{5}{3}$$

❖ *Solving compound inequalities (AND)*

The graph of a compound inequality with the word "**and**" denotes the intersection of the graph of the inequalities. A number is a solution of the compound inequality, if the number is **a solution of both inequalities.**

In the examples below, when solving, we draw a number line and express the solutions as a set-builder notation and interval notation in order to become familiar with both notations.

The set-builder notation is read as follows:

$$\{ x \mid x \le -3 \}$$

The set of   all x   such that   x is less than −3

Example 1

Solve the compound inequalities: $x + 4 < 1$ **and** $x + 2 < 3$

To solve a compound inequality, solve each inequality separately, and then graph.

$x + 4 < 1$ **and** $x + 2 < 3$

$x < 1 - 4$ (Subtract 4 from both sides.)   **and** $x + 2 < 3 - 2$ (Subtract 2 from both sides.)

$x < -3$ **and** $x < 1$

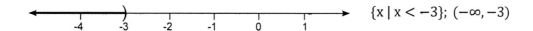

$\{x \mid x < -3\};\ (-\infty, -3)$

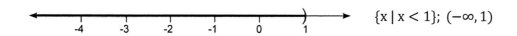

$\{x \mid x < 1\};\ (-\infty, 1)$

The solution set is $\{x \mid x < -3\}$; $(-\infty, -3)$

Example 2

Solve the compound inequalities: $2x \geq 8$ **and** $2x - 1 \leq 5$

To solve a compound inequality, solve each inequality separately, and then graph.

$2x \geq 8$ **and** $2x - 1 \leq 5$

$x \geq 4$ (Divide by 2.)   **and** $2x \leq 5 + 1$ (Add 1 to both sides.)

                                                    $2x \leq 6$ (Combine like terms.)

                                                    $x \leq 3$ (Divide by 2.)

$\{x \mid x \geq 4\};\ [4, \infty)$

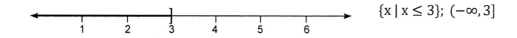

$\{x \mid x \leq 3\};\ (-\infty, 3]$

There are no solutions.

Example 3

Solve the compound inequalities (Combined form): $1 < x + 5 < 3$

In this case, the goal is to isolate x.

$1 < x + 5 < 3$

$1 - 5 < x + 5 - 5 < 3 - 5$ (Subtract 5 from all three parts of the inequality.)

$-4 < x < -2$ (Combine like terms.)

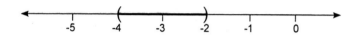

The solution set is $\{x \mid -4 < x < -2\}$; $(-4, -2)$

Example 4

Solve the compound inequalities (Combined form): $4 \le 5 - x \le 8$.

In this case, the goal is to isolate x.

$4 \le 5 - x \le 8$

$4 - 5 \le 5 - x - 5 \le 8 - 5$ (Subtract 5 from all three parts of the inequality.)

$-1 \le -x \le 3$ (Combine like terms.)

$1 \ge x \ge -3$ (Multiply or divide all three parts by -1 and switch the symbols.)

This is similar to $-3 \le x \le 1$

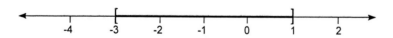

The solution set is $\{x \mid -3 \le x \le 1\}$; $[-3, 1]$

Example 5

Solve the compound inequalities (Combined form): $6 < \frac{4x+2}{3} < 10$

In this case, the goal is to isolate x. First clear the fraction.

$6 < \frac{4x+2}{3} < 10$

$6(3) < \frac{3(4x+2)}{3} < 10(3)$ (Multiply all three parts of the inequality by 3.)

$18 < 4x + 2 < 30$ (Clear the fraction.)

$18 - 2 < 4x + 2 - 2 < 30 - 2$ (Subtract 2 from all three parts of the inequality.)

$16 < 4x < 28$ (Combine like terms.)

$4 < x < 7$ (Divide all three parts by 4.)

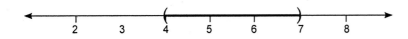

The solution set is $\{x \mid 4 < x < 7 \}$; $(4, 7)$

❖ *Solving compound inequalities (OR)*

The graph of a compound inequality with the word "**or**" denotes the union of the graphs of the inequalities. A number is a solution of the compound inequality, if the number is **a solution of at least one of the inequalities.**

Example 6

Solve the compound inequalities: $7x - 5 \leq 6$ **or** $x + 2 \geq 5$

To solve a compound inequality, solve each inequality separately, and then graph.

$7x - 5 \leq 6$	**or**	$x + 2 \geq 5$
$7x \leq 6 + 5$ (Add 5 to both sides.)	**or**	$x \geq 5 - 2$ (Subtract 2 from both sides.)
$7x \leq 11$ (Combine like terms)	**or**	$x \geq 3$ (Combine like terms.)

$x \leq \frac{11}{7}$ (Divide by 7.)

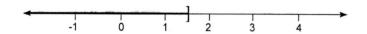

$\{x \mid x \leq \frac{11}{7}\};\ (-\infty, \frac{11}{7}]$

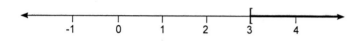

$\{x \mid x \geq 3\};\ [3, \infty)$

The solution set is $\left\{ x \mid x \leq \frac{11}{7} \text{ or } x \geq 3 \right\};\ \left(-\infty, \frac{11}{7}\right] \cup [3, \infty)$

Example 7

Solve the compound inequalities: $x - 7 < -5$ **or** $-x - 1 < 0$

To solve a compound inequality, solve each inequality separately, and then graph.

$x - 7 < -5$	**or**	$-x - 1 < 0$
$x < -5 + 7$ (Add 7 to both sides.)	**or**	$-x < 0 + 1$ (Add 1 to both sides.)
$x < 2$ (Combine like terms.)	**or**	$-x < 1$ (Combine like terms.)
		$x > -1$ (Multiply or divide by -1 and switch the symbol.)

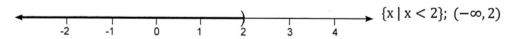

$\{x \mid x < 2\};\ (-\infty, 2)$

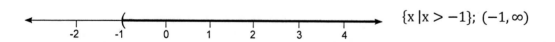

$\{x \mid x > -1\};\ (-1, \infty)$

The solution set is $\{x \mid x < 2 \text{ or } x > -1 \};\ (-\infty, \infty);$ all real numbers

Section A.2 – Compound inequalities

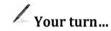 **Your turn...**

Solve the compound inequalities. Write the answers in set-builder and interval notations.

1. $x + 5 < 3$ and $x - 3 < -4$

2. $3x \geq 15$ and $4x - 3 \leq 9$

3. $-2 < x + 6 < 11$

4. $8 \leq 6 - x \leq 13$

5. $1 < \dfrac{2x+1}{5} < 4$

6. $2x - 3 \leq 10$ or $x + 7 \geq 16$

7. $x - 9 < -1$ or $-x - 5 < 0$

8. $x < 2$ or $x < 5$

For more practice, access the online homework. See your syllabus for details.

CHAPTER 1
LINEAR EQUATIONS, RELATIONS, AND FUNCTIONS

Section 1.1 – Linear Equations

- ❖ Introducing linear equations
- ❖ Graphing linear equations – intercept method
- ❖ Graphing linear equations – special cases

❖ *Introducing linear equations*

In this section, linear equations with two variables are discussed. One form of a linear equation with two variables is the standard form: Ax + By = C.

A few examples are listed below:

$$Ax + By = C$$

$$2x + 4y = 10$$

$$-5x + 7y = -3$$

$$\frac{1}{2}x + \frac{5}{3}y = 6$$

Graphically, a linear equation is a line. On a line, there is an infinite number of points or solutions. A point is on the line or is a solution of the linear equation when the coordinates of the point are plugged into the equation, and one side of the equation is equal to the other side.

Example 1

The linear equation $2x + 4y = 10$ is given. Is the point $(-1, 3)$ a solution of the equation?

Replace x with -1 and y with 3 in the given equation and simplify. If one side is equal to the other side, the point is a solution of the given equation.

$$2(-1) + 4(3) = 10$$

$$-2 + 12 = 10$$

$$10 = 10$$

The point $(-1, 3)$ is a solution of the equation.

Example 2

The linear equation $-5x + y = -3$ is given. Is the point $(-2, -5)$ a solution of the equation?

Replace x with -2 and y with -5 in the given equation. If one side is equal to the other side, the point is a solution of the given equation.

$-5(-2) + (-5) = -3$

$10 - 5 = -3$

$5 \neq -3$

The point $(-2, -5)$ is **not** a solution of the equation.

❖ *Graphing linear equations – intercept method*

Many methods can be used to graph a linear equation with two variables. One method that is first discussed in this section is the **intercept method**. When using this method, replace x with 0 and solve for y. Then replace y with 0 and solve for x. The result is two points or two intercepts $(0, y)$ and $(x, 0)$. The x-intercept is the point at which the line crosses the x-axis, where the y-value equals 0. The y-intercept is the point at which the line crosses the y-axis, where the x-value equals 0.

Example 3

Graph the linear equation $3x + 6y = 12$ using the intercept method.

To find the y-intercept

$x = 0$

$3(0) + 6y = 12$

$6y = 12$ (Divide both sides by 6.)

$y = 2$

The y-intercept is $(0, 2)$.

To find the x-intercept

$y = 0$

$3x + 6(0) = 12$

$3x = 12$ (Divide both sides by 3.)

$x = 4$

The x-intercept is $(4, 0)$.

X	Y
0	2
4	0

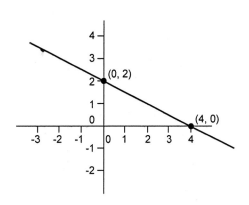

Example 4

Graph the linear equation $-3x - 5y = 15$ using the intercept method.

To find the y-intercept

$x = 0$

$-3(0) - 5y = 15$

$-5y = 15$ (Divide both sides by -5.)

$y = -3$

The y-intercept is $(0, -3)$.

To find the x-intercept

$y = 0$

$-3x - 5(0) = 15$

$-3x = 15$ (Divide both sides by -3.)

$x = -5$

The x-intercept is $(-5, 0)$.

X	Y
0	−3
−5	0

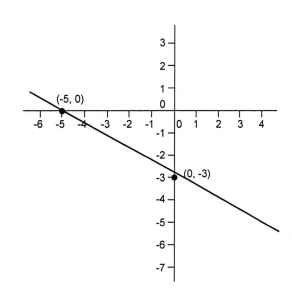

Examples 5

Laurie owes the business office $500 dollars. After discussing her situation with a staff member, a payment arrangement was made. She has to pay $25 every month until the debt is paid off. The linear equation that best represents this scenario is given by $y = 25x - 500$, x is the number of months and y the amount of money owed.

    a) How much does she still owe 10 days after the arrangement was made?
    b) How many months will it take to pay off the debt?

Solution

    a) It has not been a month yet, so she has not made her first payment. In this case, the number of month is 0 or $x = 0$.

$y = 25x - 500$

$y = 25(0) - 500$ (Replace x with 0.)

$y = -500$

$(0, -500)$

After 10 days or 0 month, she still owes $500.

    b) To pay off the loan means that the amount owed or y has to be 0.

$y = 25x - 500$

$0 = 25x - 500$ (Replace y with 0.)

$500 = 25x$ (Add 500 to both sides.)

$\dfrac{500}{25} = x$ (Divide both sides by 25.)

$20 = x$

$(20, 0)$

It will take Laurie 20 months to pay off the debt or for the amount of money owed to be zero.

    ❖ *Graphing linear equations – special cases*

Graphically, a linear equation with both variables such as the ones discussed earlier looks like a slanted or oblique line either going up or down. However, there are special cases. A linear equation can have only one of the variables and they are in the form of $x = a$ or $y = b$. In these cases, $x = a$ is a vertical line and $y = b$ is a horizontal line.

Example 6

Graph the linear equations: $x = 2$ and $x = -3$.

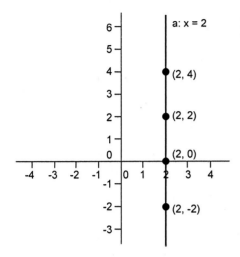 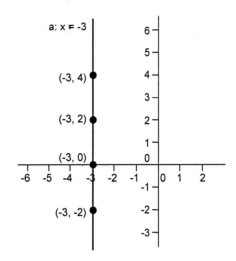

Note: All the x-values are 2 for the first graph and -3 for the second graph.

Example 7

Graph the linear equations: $y = -2$ and $y = 4$.

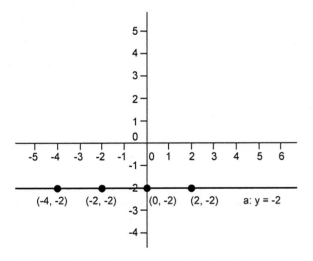 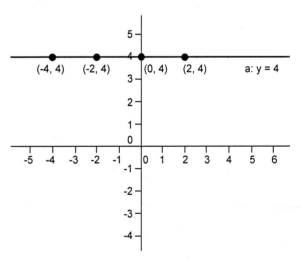

Note: All the y-values are -2 for the first graph and 4 for the second graph.

Section 1.1 – Linear equations

 **Your turn...**

Determine if a point is a solution of a linear equation.

1. The linear equation $2x + 4y = 10$ is given. Is the point $(-1, 2)$ a solution of the equation?
2. The linear equation $x - 4y = 10$ is given. Is the point $(3, -2)$ a solution of the equation?
3. The linear equation $-5x + y = -3$ is given. Is the point $(2, 7)$ a solution of the equation?

Graph the equations using the intercept method.

4. $8x - 4y = 2$
5. $x - 3y = 6$
6. $6x - 12y = 24$
7. $2x - 5y = -10$

Graph the equations.

8. $x = -2$
9. $x = 4.$
10. $y = 2$
11. $y = -3.$

Solve an application of intercepts.

12. Peter owes the business office $840 dollars. After discussing his situation with a staff member, a payment arrangement was made. He has to pay $40 every month until the debt is paid off. The linear equation that best represents this scenario is given by $y = 40x - 840$, x is the number of months and y the amount of money owed.
    a) How much does he still owe 15 days after the arrangement was made?
    b) How many months will it take to pay off the debt?

For more practice, access the online homework. See your syllabus for details.

Section 1.2 – Slope of a Line

- ❖ Finding the slope of a line given a graph
- ❖ Finding the slope of a line given two points
- ❖ Finding the slope of a line given an equation
- ❖ Using the slope to identify parallel and perpendicular lines
- ❖ Using the slope to graph a linear equation

❖ *Finding the slope of a line given a graph*

The slope of a line can be described as the incline or steepness of a line or how the graph changes. Calculating the slope simply means finding the incline of a line. The letter commonly used for slope is *m*, or moving along the axis. There are many methods that can be used to find or to identify the slope. **Three methods** are discussed in this section and they are:

Method # 1: Counting Slope

Method # 2: Slope Formula

Method # 3: Slope-Intercept Form

Method #1 - Given a graph, the slope is:

$$m = \frac{rise}{run} \ or \ \frac{vertical\ distance}{horizontal\ distance}$$

Example 1

Find the slope using the graph below.

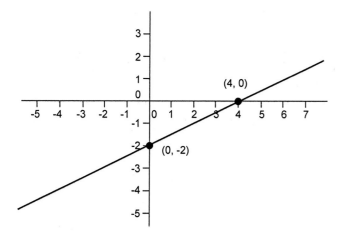

Starting on the y-axis with the y-intercept $(0, -2)$ and moving toward the x-intercept $(4, 0)$, the rise or vertical distance is 2 or 2 units up or positive direction. The run or horizontal distance is 4 or 4 units to the right or positive direction. The slope is $\frac{2}{4}$ or $\frac{1}{2}$.

Starting on the x-axis with the x-intercept $(4, 0)$ and moving toward the y-intercept $(0, -2)$, the run or horizontal distance is -4 or 4 units to the left or negative direction. The rise or vertical distance is -2 or 2 units down or negative direction. The slope is $\frac{-2}{-4}$ or $\frac{1}{2}$.

Example 2

Find the slope using the graph below.

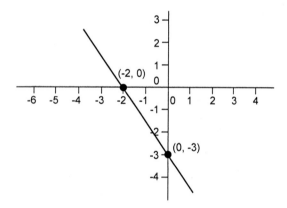

Starting on the y-axis with the y-intercept $(0, -3)$ and moving toward the x-intercept $(-2, 0)$, the rise or vertical distance is 3 or 3 units up or positive direction. The run or horizontal distance is -2 or 2 units to the left or negative direction. The slope is $\frac{3}{-2}$ or $-\frac{3}{2}$

Starting on the x-axis with the x-intercept $(-2, 0)$ and moving toward the y-intercept $(0, -3)$, the run or horizontal distance is 2 or 2 units to the right or positive direction. The rise or vertical distance is -3 or 3 units down or negative direction. The slope is $\frac{-3}{2}$ or $-\frac{3}{2}$.

❖ *Finding the slope of a line given two points*

Method # 2 - To solve for the slope given 2 points, use the formula below:

$$m = \frac{y_2 - y_1}{x_2 - x_1}$$

Example 3

Find the slope of the line passing through the points $(5, 3)$ and $(-1, 3)$.

$$(x_1, y_1) \text{ and } (x_2, y_2)$$

Using the formula $m = \frac{y_2 - y_1}{x_2 - x_1}$

$$m = \frac{3 - 3}{-1 - 5} = \frac{0}{-6} = 0$$

The slope is zero. The line is horizontal.

Example 4

Find the slope of the line passing through the points $(-4, 1)$ and $(-4, 7)$.

$$(x_1, y_1) \text{ and } (x_2, y_2)$$

Using the formula $m = \frac{y_2 - y_1}{x_2 - x_1}$

$$m = \frac{7 - 1}{-4 - (-4)} = \frac{6}{0}$$

The slope is undefined. The line is vertical.

Example 5

Find the slope of the line passing through the points $(-3, 2)$ and $(4, 5)$.

$$(x_1, y_1) \text{ and } (x_2, y_2)$$

Using the formula $m = \frac{y_2 - y_1}{x_2 - x_1}$

$$m = \frac{5 - 2}{4 - (-3)} = \frac{3}{7}$$

The slope is $\frac{3}{7}$. A positive slope means that from left to right the line goes up.

Example 6

Find the slope of the line passing through the points $(-3, 2)$ and $(8, -5)$.

$$(x_1, y_1) \text{ and } (x_2, y_2)$$

Using the formula $m = \frac{y_2 - y_1}{x_2 - x_1}$

$$m = \frac{-5 - 2}{8 - (-3)} = \frac{-7}{11}$$

The slope is $\frac{-7}{11}$. A negative slope means that from left to right the line goes down.

Example 7

On one of the Valencia Campuses, the number of students enrolled in 2005 was 14,985. In 2010, this number increased to 17,982. Find the slope. Explain the meaning of the slope.

Let us express the data as a set of points: (2005, 14985) and (2010, 17982).

$$(x_1, y_1) \quad \text{and} \quad (x_2, y_2)$$

Using the formula $m = \frac{y_2 - y_1}{x_2 - x_1}$

$$m = \frac{17982 - 14985}{2010 - 2005} = \frac{2997}{5} = 599.4$$

The slope is approximately 599. It means that the number of students enrolled at Valencia for this particular campus increased by 599 every year from 2005 to 2010.

Example 8

In a certain university, the number of students who majored in mathematics in 1988 was 130. By 2006, only 55 students majored in mathematics. Find the slope. Explain the meaning of the slope.

Let us express the data as a set of points: (1988, 130) and (2006, 55).

$$(x_1, y_1) \quad \text{and} \quad (x_2, y_2)$$

Using the formula $m = \frac{y_2 - y_1}{x_2 - x_1}$

$$m = \frac{55 - 130}{2006 - 1988} = \frac{-75}{18} = -4.17 \qquad (\text{Approximately} - 4)$$

The slope is -4. It means that the number of students who majored in mathematics decreased by 4 every year.

❖ *Finding the slope of a line given an equation*

Method # 3 – Another form of the linear equation is the slope-intercept form $y = mx + b$. The slope is the coefficient of x and the y-intercept is $(0, b)$. A few examples on how to **identify the slope** are listed below:

$$y = \frac{2}{3}x + 5; \text{ the slope is } \frac{2}{3}.$$

$$y = -5x + 7; \text{ the slope is } -5.$$

$$y = -\frac{1}{3}x - \frac{5}{3}; \text{ the slope is } -\frac{1}{3}.$$

Note: The equation must be in the form $y = mx + b$.

Example 9

Find the slope and y-intercept of the equation $y = -x + \frac{3}{5}$.

Remember that the slope is identified as the coefficient of x when the equation is in the slope-intercept form $y = mx + b$.

For the equation $y = -x + \frac{3}{5}$, the slope is -1 and the $y-$ intercept is $(0, \frac{3}{5})$.

Example 10

Find the slope and y-intercept of the equation $3x + 4y = 7$.

Remember that the slope is identified as the coefficient of x when the equation is in the slope-intercept form $y = mx + b$. We need to put this equation in the slope-intercept form by solving for y.

$3x + 4y = 7$

$4y = -3x + 7$ (Subtract $3x$ from both sides.)

$y = \frac{-3}{4}x + \frac{7}{4}$ (Divide both sides by 4.)

$y = \frac{-3}{4}x + \frac{7}{4}$

The slope is $-\frac{3}{4}$ and the $y-$ intercept is $\left(0, \frac{7}{4}\right)$.

Example 11

Find the slope and y-intercept of the equation $\frac{1}{4}y - x = 2$.

Remember that the slope is identified as the coefficient of x when the equation is in the slope-intercept form $y = mx + b$. We need to put this equation in the slope-intercept form by solving for y.

$\frac{1}{4}y - x = 2$

$4\left(\frac{1}{4}\right)y - 4(x) = 4(2)$ (Multiply every term by 4 to clear the fraction.)

$y - 4x = 8$ (Add $4x$ to both sides.)

$y = 4x + 8$

The slope is 4 and the $y - $ intercept is $(0, 8)$.

❖ *Using the slope to identify parallel and perpendicular lines*

The slope can be used to find out if two lines are parallel, perpendicular, or neither. **Parallel** lines have the same slope. **Perpendicular** lines have negative reciprocal slopes ($\frac{2}{3}$ and $-\frac{3}{2}$). In other words, multiply the two slopes, if the result is -1, the lines are perpendicular. If the slopes are not the same and the product of the two slopes is not -1, then the lines are **neither** parallel nor perpendicular.

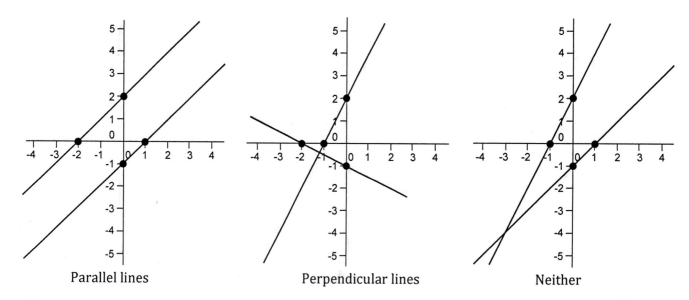

Parallel lines   Perpendicular lines   Neither

Example 12

Determine if the lines are parallel, perpendicular, or neither.

Line 1: $y = -3x + 7$; Line 2: $y = \frac{1}{3}x - 4$

The equations are already in slope-intercept form, so it is easier to identify the slopes.

The slope for line 1 is -3. The slope for line 2 is $\frac{1}{3}$.

The lines are not parallel because they do not have the same slope ($-3 \neq \frac{1}{3}$).

The lines are **perpendicular** because -3 is the negative reciprocal of $\frac{1}{3}$. Also, when the two slopes are multiplied, the result is -1 or $[-3\left(\frac{1}{3}\right) = -1]$.

Example 13

Determine if the lines are parallel, perpendicular, or neither.

Line 1: $y = 2x + \frac{1}{2}$;   Line 2: $-6x + 3y = 15$

The slope for line 1 is 2.

Line 2 is not in slope–intercept form. Solve the equation of line 2 for y, so that the slopes can be compared.

$-6x + 3y = 15$

$3y = 6x + 15$ (Add $6x$ to both sides.)

$y = \frac{6}{3}x + \frac{15}{3}$ (Divide both sides by 3 and simplify.)

$y = 2x + 5$

The slope for line 2 is also 2.

The lines are **parallel** because they have the same slope.

Example 14

Determine if the lines are parallel, perpendicular, or neither.

Line 1: $-4x + y = -2$;   Line 2: $-x + 4y = 3$

The equations are not in slope–intercept form. Solve both equations for y so that the slopes can be compared.

Solve equation 1 for y.

$-4x + y = -2$

$y = 4x - 2$ (Add $4x$ to both sides.)

The slope is 4.

Solve equation 2 for y.

$-x + 4y = 3$

$4y = x + 3$ (Add x to both sides.)

$y = \frac{x}{4} + \frac{3}{4}$ (Divide both sides by 4.)

$y = \frac{1}{4} x + \frac{3}{4}$

The slope is $\frac{1}{4}$.

The lines are not parallel because the slopes are not the same.

The lines are not perpendicular because the product of the slopes is not -1 or one slope is not the negative reciprocal of the other slope.

In this case, the answer is **neither** parallel nor perpendicular.

❖ *Using the slope to graph a linear equation*

Recall that in the previous section, the intercepts were used to graph a linear equation. The **slope and y-intercept** can also be used to graph a linear equation. To do so, plot the point or y-intercept. From the point, use the slope; rise (move up or down) over run (left or right) to obtain another point. Then connect the points to graph the linear equation.

Example 15

Graph the line $y = \frac{3}{5}x - 2$ using the slope and y-intercept.

Since the equation is in slope-intercept form, it is easier to identify the slope and y-intercept. The slope is $\frac{3}{5} = \frac{Rise}{Run}$ and the $y -$ intercept is $(0, -2)$.

From the point $(0, -2)$, move up 3 units and right 5 units, to get another point and to graph the linear equation.

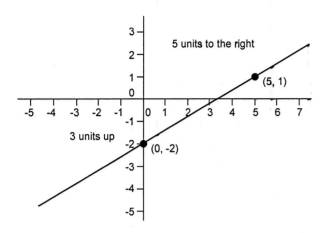

Example 16

Graph the line $4x + 3y = 9$ using the slope and y-intercept.

First, the equation needs to be in the slope-intercept form. Then identify the slope and y-intercept and use them to graph the equation.

$4x + 3y = 9$

$3y = -4x + 9$ (Subtract $4x$ from both sides.)

$y = -\frac{4}{3}x + \frac{9}{3}$ (Divide both sides by 3 and simplify.)

$y = -\frac{4}{3}x + 3$

The slope can either be $\frac{-4}{3}$ or $\frac{4}{-3}$ and the $y-$ intercept is $(0, 3)$.

From the ordered pair $(0, 3)$, if using the slope $\frac{-4}{3}$, move down 4 units and right 3 units. If using the slope $\frac{4}{-3}$, move up 4 units and left 3 units, to get another point and to graph the linear equation. The graph below shows the first option.

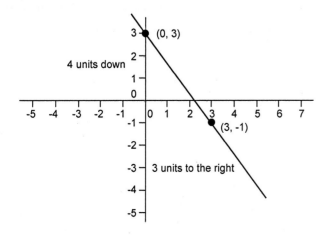

Section 1.2 – Slope of a line

Your turn...

1. Find the slope of the graph below using method # 1, counting slope.

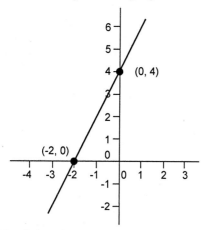

2. Find the slope of the graph below using method # 1, counting slope.

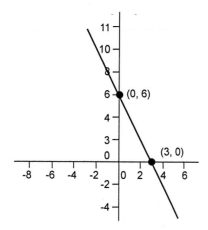

Find the slope of the line passing through the points using the slope formula.

3. $(5, 2)$ and $(-1, 2)$

4. $(-3, 1)$ and $(-3, 7)$

5. $(-1, 2)$ and $(4, 5)$

6. $(-3, 3)$ and $(8, -6)$

7. In a certain university, the number of students who majored in mathematics in 1989 was 180. By 2007, only 65 students majored in mathematics. What is the slope? What is the meaning of the slope in the context of this problem?

8. On one of the Valencia Campuses, the number of students enrolled in 2008 was 35,085. In 2012, this number increased to 38,653. Find the slope. Explain the meaning of the slope.

Find the slope and y-intercept of the equations.

9. $y = -2x + \frac{2}{5}$

10. $5x + 4y = 9$

11. $\frac{1}{3}y - x = 6$

Determine if the lines are parallel, perpendicular, or neither.

12. $y = -2x + 1$; $y = \frac{1}{2}x - 4$

13. $y = 2x + \frac{1}{9}$; $-12x + 6y = 30$

14. $-7x + y = -3$; $-x + 7y = 5$

Graph the linear equations using the slope and y-intercept.

15. $y = \frac{1}{2}x - 4$

16. $4x + 5y = 10$

For more practice, access the online homework. See your syllabus for details.

Section 1.3 – Equations of a Line

- ❖ Finding an equation in slope-intercept form given a slope and a point
- ❖ Finding an equation given two points
- ❖ Finding an equation of a line parallel to another equation
- ❖ Finding an equation of a line perpendicular to another equation
- ❖ Solving applications of linear equations

❖ *Finding an equation in slope-intercept form given a slope and a point*

In the previous section, the linear equation is used to answer questions about slope, y-intercept, parallel lines, and perpendicular lines. In this section, given specific information, we will find the linear equation. When finding the equation of the line, we need to have a slope and at most one point. We can use the **point-slope form of the equation** below.

$$y - y_1 = m(x - x_1), \qquad \text{where } (x_1, y_1) \text{ is the point and } m \text{ is the slope.}$$

Given a slope and a point, replace x_1, y_1, and m in the above equation, simplify, and express the final equation in slope-intercept form.

Example 1

Find an equation of the line having slope 2 and passing through the point $(-1, 7)$ or (x_1, y_1). Express the final answer in slope-intercept form.

Using the point-slope form of the equation, $y - y_1 = m(x - x_1)$, plug in the slope and the point, and simplify.

$y - y_1 = m(x - x_1)$

$y - 7 = 2[x - (-1)]$ (Replace m with 2, x_1 with -1, and y_1 with 7.)

$y - 7 = 2(x + 1)$ (The 2 negative signs become positive.)

$y - 7 = 2x + 2$ (Distribute the 2.)

$y = 2x + 2 + 7$ (Add 7 to both sides.)

$y = 2x + 9$

Example 2

Find an equation of the line having the slope $-\frac{2}{7}$ and point $(4, 9)$ or (x_1, y_1). Express the final answer in slope-intercept form.

Using the point-slope form of the equation, $y - y_1 = m(x - x_1)$, plug in the slope and the point, and simplify.

$$y - y_1 = m(x - x_1)$$

$$y - 9 = -\frac{2}{7}(x - 4) \qquad \text{(Replace } m \text{ with } -\frac{2}{7}, x_1 \text{ with 4, and } y_1 \text{ with 9.)}$$

$$y - 9 = -\frac{2}{7}x + \frac{8}{7} \qquad \text{(Distribute the } -\frac{2}{7}.\text{)}$$

$$y = -\frac{2}{7}x + \frac{8}{7} + 9 \qquad \text{(Add 9 to both sides and simplify } \frac{8}{7} + \frac{9}{1} = \frac{8}{7} + \frac{63}{7} = \frac{71}{7}.\text{)}$$

$$y = -\frac{2}{7}x + \frac{71}{7}$$

❖ **Finding an equation given two points**

Given 2 points (x_1, y_1) and (x_2, y_2), first find the slope using the slope formula $m = \frac{y_2 - y_1}{x_2 - x_1}$.

Use the slope found, and **either one** of the points, and replace them in the point-slope form of the equation. Simplify and express the final equation in slope-intercept form.

Example 3

Find an equation of the line passing through the points $(1, 2)$ and $(4, 8)$ or (x_1, y_1) and (x_2, y_2). Express the final answer in slope-intercept form.

Before finding the equation of the line, we need to find the slope.

Using the formula $m = \frac{y_2 - y_1}{x_2 - x_1}$, let us solve for the slope.

$$m = \frac{8 - 2}{4 - 1} = \frac{6}{3} = 2$$

Using the point-slope form of the equation, $y - y_1 = m(x - x_1)$, plug in the slope, one of the points, and simplify. **It does not matter which point is used**, the equation will be the same at the end. Using the first point $(1, 2)$ as (x_1, y_1) and slope 2, we have:

$$y - y_1 = m(x - x_1)$$

$$y - 2 = 2(x - 1) \qquad \text{(Replace } m \text{ with 2, } x_1 \text{ with 1, and } y_1 \text{ with 2.)}$$

$$y - 2 = 2x - 2 \qquad \text{(Distribute the 2.)}$$

$$y = 2x - 2 + 2 \rightarrow y = 2x \qquad \text{(Add 2 to both sides and simplify.)}$$

Example 4

Find an equation of the line passing through the points $(-1, 2)$ and $(4, 5)$ or (x_1, y_1) and (x_2, y_2). Express the final answer in slope-intercept form.

Before finding the equation of the line, we need to find the slope.

Using the formula $m = \frac{y_2 - y_1}{x_2 - x_1}$, let us solve for the slope.

$$m = \frac{5 - 2}{4 - (-1)} = \frac{3}{5}$$

Using the point-slope form of the equation, $y - y_1 = m(x - x_1)$, plug in the slope, one of the points, and simplify. **It does not matter which point is used**, the equation will be the same at the end. Using the second point $(4, 5)$ as (x_1, y_1) and slope $\frac{3}{5}$, we have :

$$y - y_1 = m(x - x_1)$$

$$y - 5 = \frac{3}{5}(x - 4) \qquad \text{(Replace } m \text{ with } \frac{3}{5}, x_1 \text{ with 4, and } y_1 \text{ with 5.)}$$

$$y - 5 = \frac{3}{5}x - \frac{12}{5} \qquad \text{(Distribute the } \frac{3}{5}.)$$

$$y = \frac{3}{5}x - \frac{12}{5} + 5 \qquad \text{(Add 5 to both sides and simplify } -\frac{12}{5} + \frac{5}{1} = -\frac{12}{5} + \frac{25}{5} = \frac{13}{5}.)$$

$$y = \frac{3}{5}x + \frac{13}{5}$$

❖ *Finding an equation of a line parallel to another equation (or finding an equation of a line given a point and parallel line)*

Given a point and an equation, find the slope by solving the given equation for y or putting the equation in the slope-intercept form if necessary. The coefficient of x is the slope. Then use **the same slope found**, the given point, and replace them in the point-slope form of the equation. Simplify and express the final equation in slope-intercept form.

Note: The slope-intercept form of the equation can be used as well. For the sake of consistency, only the point-slope form of the equation is used in this book.

Example 5

Find an equation of the line in slope-intercept form containing the point $(-3, -2)$ and **parallel** to the equation or line $y = -7x + 2$.

Before finding the equation of the line, we need to find the slope. We can get the slope by using the given equation $y = -7x + 2$. Since the given equation is in the slope-intercept form, we can identify the slope as $m = -7$.

The key word is **parallel**. Since the lines are parallel they should have the **same slopes**. The **slope** is -7. The point given is $(-3, -2)$ or (x_1, y_1).

$y - y_1 = m(x - x_1)$

$y - (-2) = -7[x - (-3)]$ (Replace m with -7, x_1 with -3, and y_1 with -2.)

$y + 2 = -7(x + 3)$ (The 2 negative signs become positive.)

$y + 2 = -7x - 21$ (Distribute the -7.)

$y = -7x - 21 - 2$ (Subtract 2 from both sides.)

$y = -7x - 23$

Example 6

Find an equation of the line in slope-intercept form containing the point $(4, 5)$ and **parallel** to the equation or line $x + 5y = 20$.

Before finding the equation of the line, we need to find the slope. We can get the slope by using the given equation $x + 5y = 20$. Since the given equation is not in the slope-intercept form, put it in that form to find the slope.

$x + 5y = 20$

$5y = -x + 20$ (Subtract x from both sides.)

$y = -\frac{x}{5} + \frac{20}{5}$ (Divide both sides by 5.)

$y = -\frac{1}{5}x + \frac{20}{5}$ or $y = -\frac{1}{5}x + 4$

We can now identify the slope as $m = -\frac{1}{5}$.

The key word is **parallel**. When putting together the equation, we use the **same slope** which is $-\frac{1}{5}$. The point given is $(4, 5)$ or (x_1, y_1).

$y - y_1 = m(x - x_1)$

$y - 5 = -\frac{1}{5}(x - 4)$ (Replace m with $-\frac{1}{5}$, x_1 with 4, and y_1 with 5.)

$y - 5 = -\frac{1}{5}x + \frac{4}{5}$ (Distribute the $-\frac{1}{5}$.)

$y = -\frac{1}{5}x + \frac{4}{5} + 5$ (Add 5 to both sides and simplify $\frac{4}{5} + \frac{5}{1} = \frac{4}{5} + \frac{25}{5} = \frac{29}{5}$.)

$y = -\frac{1}{5}x + \frac{29}{5}$

❖ *Finding an equation of a line perpendicular to another equation (or finding an equation of a line given a point and perpendicular line)*

Given a point and an equation, find the slope by solving the given equation for y or by putting the equation in the slope-intercept form if necessary. The coefficient of x is the slope. Then use the **negative reciprocal of the slope found**, the given point, and replace them in the point-slope form of the equation. Simplify and express the final equation in slope-intercept form.

Note: The slope-intercept form of the equation can be used as well. For the sake of consistency, only the point-slope form of the equation is used in this book.

Example 7

Find an equation of the line containing the point $(-3, 4)$ and **perpendicular** to the equation or line $x + 3y = 20$.

Before finding the equation of the line, we need to find the slope. We can get the slope by using the given equation $x + 3y = 20$. Since the given equation is not in the slope-intercept form, put it in that form to find the slope.

$x + 3y = 20$

$3y = -x + 20$ (Subtract x from both sides.)

$y = -\dfrac{x}{3} + \dfrac{20}{3}$ (Divide both sides by 3.)

$y = -\dfrac{1}{3}x + \dfrac{20}{3}$

We can now identify the slope as $m = -\dfrac{1}{3}$.

The key word is **perpendicular.** When putting together the equation, we use the **negative reciprocal of the slope** which is 3. The point given is $(-3, 4)$ or (x_1, y_1).

$y - y_1 = m(x - x_1)$

$y - 4 = 3[x - (-3)]$ (Replace m with 3 , x_1 with -3, and y_1 with 4.)

$y - 4 = 3(x + 3)$ (The 2 negative signs become positive.)

$y - 4 = 3x + 9$ (Distribute the 3.)

$y = 3x + 9 + 4$ (Add 4 to both sides.)

$y = 3x + 13$

Example 8

Find an equation of the line containing the point $(1, 2)$ and **perpendicular** to the equation or line $y = -7x + 2$.

Before finding the equation of the line, we need to find the slope. We can get the slope by using the given equation $y = -7x + 2$. Since the given equation is in the slope-intercept form, we can identify the slope as $m = -7$.

The key word is **perpendicular.** When putting together the equation, we use the **negative reciprocal of the slope** which is $\frac{1}{7}$. The point given is $(1, 2)$ or (x_1, y_1).

$y - y_1 = m(x - x_1)$

$y - 2 = \frac{1}{7}(x - 1)$ (Replace m with $\frac{1}{7}$, x_1 with 1, and y_1 with 2.)

$y - 2 = \frac{1}{7}x - \frac{1}{7}$ (Distribute the $\frac{1}{7}$.)

$y = \frac{1}{7}x - \frac{1}{7} + 2$ (Add 2 to both sides and simplify $-\frac{1}{7} + \frac{2}{1} = -\frac{1}{7} + \frac{14}{7} = \frac{13}{7}$.)

$y = \frac{1}{7}x + \frac{13}{7}$

❖ *Solving applications of linear equations*

Example 9

Anita calls a plumber because she is having problems with her kitchen sink. The plumber charges her a flat fee of $25 plus $12.50 per hour.

   a) Write a linear equation where y is the amount of money charged based on the amount of time x.
   b) How much will Anita pay the plumber if it takes him 3 hours to fix the problem?

Solution

   a) We use the slope-intercept form of the equation $y = mx + b$ to solve this type of problem, where b is 25 (fixed) and m is 12.50 (variable). The linear equation is $y = 12.50x + 25$.

   b) Using the above equation $y = 12.50x + 25$, we replace x with 3.

      $y = 12.50(3) + 25$

      $y = 37.50 + 25$

      $y = \$62.50$

      Anita was charged $62.50.

Example 10

Alan and Doris are no longer happy with their cable provider. Their new provider, Cable-Right, charges them $40 for a visit to their home plus $19 per hour to set up the service. It took the cable man 2 hours and 30 minutes to set up the service. How much was charged to Alan and Doris' account?

Solution

First, let us find the linear equation for this problem. Using $y = mx + b$, where b is 40 (fixed) and m is 19 (variable), we have:

$y = 19x + 40$

Using the above equation, we can find out how much they were charged.

$y = 19x + 40$

$y = 19(2.5) + 40$ (Replace x with 2.5 hours.)

$y = 47.5 + 40$

$y = \$87.50$

Alan and Doris were charged $87.50 for the set-up of their new cable service.

Example 11

It has been shown that having a student leader in the classroom helps students succeed. At Valencia College, in 2006, 42 sections or classes had Supplemental Learning leaders. In 2008, 226 sections had Supplemental Learning leaders. This shows that the program has experienced a tremendous growth. Use the given data to find an equation of the line that predicts the future growth of the program.

Solution

The data can be expressed as points (2006, 42) and (2008, 226) or (x_1, y_1) and (x_2, y_2).

Before finding the equation of the line, we need to find the slope.

Using the formula $m = \frac{y_2 - y_1}{x_2 - x_1}$, let us solve for the slope.

$$m = \frac{226 - 42}{2008 - 2006} = \frac{184}{2} = 92$$

Using the point-slope of the equation, $y - y_1 = m(x - x_1)$, plug in the slope, one of the points, and simplify. **It does not matter which point is used**, the equation will be the same at the end. Using the first point $(2006, 42)$ as (x_1, y_1) and slope 92, we have:

$y - y_1 = m(x - x_1)$

$y - 42 = 92(x - 2006)$ (Replace m with 92, x_1 with 2006, and y_1 with 42.)

$y - 42 = 92x - 184{,}552$ (Distribute the 92.)

$y = 92x - 184{,}552 + 42$ (Add 42 to both sides.)

$y = 92x - 184{,}510$

Example 12

In a certain university, the number of students who majored in Statistics in 1990 was 350. By 2005, only 50 students majored in Statistics. Find an equation of the line.

Solution

The data can be expressed as points $(1990, 350)$ and $(2005, 50)$ or (x_1, y_1) and (x_2, y_2).

Before finding the equation of the line, we need to find the slope.

Using the formula $m = \frac{y_2 - y_1}{x_2 - x_1}$, let us solve for the slope.

$$m = \frac{50 - 350}{2005 - 1990} = \frac{-300}{15} = -20$$

Using the point-slope of the equation, $y - y_1 = m(x - x_1)$, plug in the slope, one of the points, and simplify. **It does not matter which point is used**, the equation will be the same at the end. Using the first point $(2005, 50)$ as (x_1, y_1) and slope -20, we have:

$y - y_1 = m(x - x_1)$

$y - 50 = -20(x - 2005)$ (Replace m with -20, x_1 with 2005, and y_1 with 50.)

$y - 50 = -20x + 40100$ (Distribute the -20.)

$y = -20x + 40100 + 50$ (Add 50 to both sides.)

$y = -20x + 40{,}150$

Section 1.3 – Equations of a line

Your turn...

1. Find an equation of the line having a slope of 3 and point $(-1, 5)$.

2. Find an equation of the line having a slope of $-\frac{1}{7}$ and point $(2, 7)$.

3. Find an equation of the line passing through the points $(3, 4)$ and $(5, 6)$.

4. Find an equation of the line passing through the points $(-1, 3)$ and $(4, 7)$.

5. Find an equation of the line containing the point $(-1, -2)$ and **parallel** to $y = -6x + 2$.

6. Find an equation of the line containing the point $(3, 2)$ and **parallel** to $x + 2y = 20$.

7. Find an equation of the line containing the point $(-3, 5)$ and **perpendicular** to $x + 3y = 20$.

8. Find an equation of the line containing the point $(1, 2)$ and **perpendicular** to $y = -3x + 2$.

9. Ishtar calls a plumber because she is having problems with her kitchen sink. The plumber charges her a flat fee of $15 and $11.50 per hour.
   a) Write a linear equation where y is the amount of money charged based on the amount of time x.
   b) How much will Ishtar pay the plumber if it takes him 4 hours to fix the problem?

10. Lee and May are no longer happy with their cable provider. Their new provider, Cable-Right, charges them $29 for a visit to their home and $13 per hour to set up the service. It took the cable man 3 hours to set up the service. How much was charged to Lee and May's account?

11. It has been shown that having a student leader in the classroom helps students succeed. At Valencia College, in 2007, 177 sections or classes had Supplemental Learning leaders. In 2010, 320 sections had Supplemental Learning leaders. This shows that the program has experienced a tremendous growth. Use the given data to find an equation of the line that predicts the future growth of the program.

12. In a certain university, the number of students who majored in Psychology in 1995 was 850. By 2010, the number of students who majored in Psychology increased to 1,090. Find an equation of the line.

For more practice, access the online homework. See your syllabus for details.

Section 1.4 – Relations

❖ Finding the domain and range of a relation

❖ *Finding the domain and range of a relation*

A relation is sometimes expressed as a set of points. Below are examples of relations:

$$\{(1, 2), (3, 4), (5, 6), (7, 8), (9, 10)\}$$

$$\{(-3, 6), (-2, 4), (-1, -2)\}$$

$$\{(1, 28), (2, 27), (3, 25), (4, 23)\}$$

The last one can also be expressed as the table below:

Months of Spring semester, x	Students in a math class, y
January (1)	28
February (2)	27
March (3)	25
April (4)	23

The **domain** of a relation refers to all the x-values or the first coordinate of the points. To find the domain of a relation, simply list all the x-values.

The **range** of a relation refers to all the y-values or the second coordinate of the points. To find the range of a relation, simply list all the y-values.

In the examples above:

The domains are $\{1, 3, 5, 7, 9\}$; $\{-3, -2, -1\}$ and {January, February, March, April}.

The ranges are $\{2, 4, 6, 8, 10\}$; $\{-2, 4, 6\}$ and $\{23, 25, 27, 28\}$.

Example 1

Find the domain and range of this relation: $\{(-7, 49), (-6, 36), (-5, 25), (4, -16), (3, -9), (2, -4)\}$.

The domain is the list of all the x-values $\{-7, -6, -5, 2, 3, 4\}$.

The range is the list of all the y-values $\{-16, -9, -4, 25, 36, 49\}$.

Example 2

Find the domain and range of this relation: $\{(0, 4), (1, 6), (2, 5), (4, -1), (3, -9), (2, -4)\}$.

The domain is the list of all the x-values $\{0, 1, 2, 3, 4\}$.

The range is the list of all the y-values $\{-9, -4, -1, 4, 5, 6\}$.

Example 3

Find the domain and range of this relation: $\{(0, 4), (1, 6), (2, 5), (4, -1), (2, 4)\}$.

The domain is the list of all the x-values $\{0, 1, 2, 4\}$.

The range is the list of all the y-values $\{-1, 4, 5, 6\}$.

Example 4

Find the domain and range of this relation.

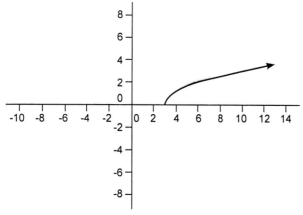

The domain includes all the x-values for which there is a y-value on the graph. For instance, $x = 2$ has no y-value. So $x = 2$ is not included in the domain.

The domain is $\{x \mid x \geq 3\}$ or $[3, \infty)$.

The range includes all the y-values on the graph for all the x-values in the domain.

The range is $\{y \mid y \geq 0\}$ or $[0, \infty)$.

Example 5

Find the domain and range of this relation.

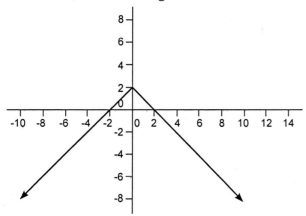

The domain includes all the *x*-values for which there is a y-value on the graph.

The domain consists of all real numbers or $(-\infty, \infty)$.

The range includes all the y-values on the graph for all the *x*-values in the domain. The highest y-value on the graph is 2. The graph does not go beyond 2.

The range is $\{y | y \le 2\}$ or $(-\infty, 2]$

Example 6

Find the domain and range of this relation.

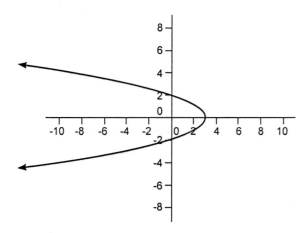

The domain includes all the x-values for which there is a y-value on the graph. For instance, x = 4 has no y-value. So x = 4 is not included in the domain. The domain is $\{x | x \le 3\}$ or $(-\infty, 3]$.

The range includes all the y-values on the graph for all the *x*-values in the domain.

The range consists of all real numbers or $(-\infty, \infty)$.

Section 1.4 – Relations

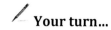 **Your turn...**

Find the domain and range for each relation.

1. $\{(-7, 9), (-6, 6), (-5, 5), (4, -6), (0, -9), (2, -1)\}$
2. $\{(0, 2), (1, 5), (2, 6), (4, -2), (3, -7), (2, -3)\}$
3. $\{(0, 5), (1, 6), (2, 3), (5, -1), (2, 5)\}$

4. Find the domain and range of this relation.

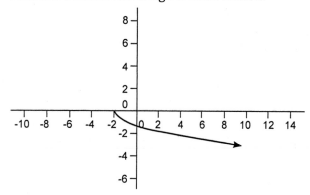

5. Find the domain and range of this relation.

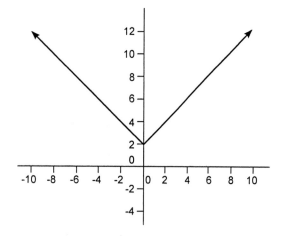

For more practice, access the online homework. See your syllabus for details.

Section 1.5 – Functions

- ❖ Introducing functions
  - o Relations
  - o Graph
- ❖ Evaluating functions
- ❖ Finding the domain of functions
- ❖ Graphing functions

❖ *Introducing functions*

A **function** has exactly one y-value for every x-value. A relation is said to be a function if every x-value has exactly one y-value. Therefore, a relation is **not** a function if one x-value **has two or more** corresponding y-values.

If the relation is expressed graphically, use the vertical line test to determine if the graph is that of a function. A vertical line drawn **anywhere** on the graph should **always** intersect the graph at only one point. The graph is **not** a function if the line intersects the graph (not the $x - axis$) at **two or more points**.

In this section, we also introduce the function notation and what it means. Looking at a linear equation, it can be expressed with function notation:

$$y = 2x + \frac{1}{7}$$

$$f(x) = 2x + \frac{1}{7}$$

"$f(x)$" $\rightarrow f$ of x or f is a function of x.

$$g(x) = 2x + \frac{1}{7}$$

"$g(x)$" $\rightarrow$ g of x or g is a function of x.

An example of V as a function t is $V(t) = 3t - 11$.

An example of H as a function a is $H(a) = -\frac{1}{2} a^2 + 5a$

○ *Relations - Determine if a relation is a function.*

Example 1

Is this relation a function? $\{(-7, 49), (-6, 36), (-5, 25)\}$

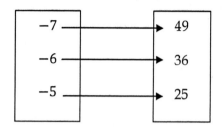

Yes. Every x has one y-value.

Example 2

Is this relation a function? $\{(-4, -9), (0, 4), (1, 4), (2, 5)\}$

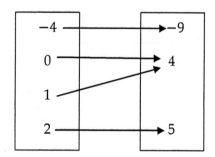

Yes. Every x has one y-value. It just happened that one x shares the same y-value.

Example 3

Is this relation a function? $\{(0, 4), (1, 6), (3, 8), (3, 10)\}$

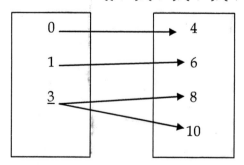

No. The x-value 3 has two y-values: 8 and 10.

    ○   *Graph - Determine if the graph shown is the graph of a function.*

Example 4

Is this the graph of a function?

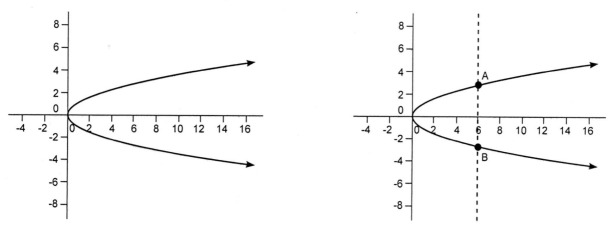

Using the vertical line test, the answer is no. The vertical line drawn on the graph intersects it at 2 places or 2 points (A and B).

Example 5

Is this the graph of a function?

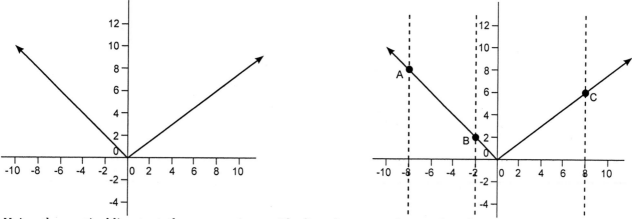

Using the vertical line test, the answer is yes. The line drawn on the graph only intersects it at one point (A or B or C).

Example 6

Is this the graph of a function?

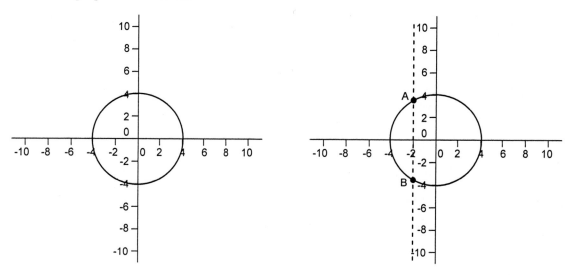

Using the vertical line test, the answer is no. The vertical line drawn on the graph intersects it at 2 places or 2 points (A and B).

❖ *Evaluating functions*

Evaluating functions mean finding the y-value given an x-value or vice versa. Look at the example below.

$M(x) = -15x + 8$

$M(3) = -15(3) + 8$

$M(3) = -37$

In the above example, given $x = 3$, the value of the function or y is -37 or $M(3) = -37$. Expressed as a point, we have $(3, -37)$.

Example 7

The function $W(x) = -9x + 4$ is given. Find $W(-2)$.

Replace x with -2.

$W(-2) = -9(-2) + 4$

$W(-2) = 18 + 4$

$W(-2) = 22$

This can also be expressed as $x = -2$ and $y = 22$ or $(-2, 22)$.

Example 8

The function $P(x) = -9x^2 + 3x - 4$ is given. Find $P(-3)$.

Replace x with -3.

$P(-3) = -9(-3)^2 + 3(-3) - 4$

$P(-3) = -9(9) - 9 - 4$

$P(-3) = -81 - 9 - 4$

$P(-3) = -94$

This can also be expressed as $x = -3$ and $y = -94$ or $(-3, -94)$.

Example 9

The function $R(x) = 5x - 4$ is given. Find $R\left(\frac{1}{5}\right)$.

Replace x with $\frac{1}{5}$.

$R\left(\frac{1}{5}\right) = 5\left(\frac{1}{5}\right) - 4$

$R\left(\frac{1}{5}\right) = 1 - 4$

$R\left(\frac{1}{5}\right) = -3$

This can also be expressed as $x = \frac{1}{5}$ and $y = -3$ or $\left(\frac{1}{5}, -3\right)$.

Example 10

The function $R(x) = 5x + 3$ is given. Find $R\left(\frac{1}{4}\right)$.

Replace x with $\frac{1}{4}$.

$R\left(\frac{1}{4}\right) = 5\left(\frac{1}{4}\right) + 3$

$R\left(\frac{1}{4}\right) = \frac{5}{4} + 3 = \frac{5}{4} + \frac{3}{1} = \frac{5}{4} + \frac{12}{4} = \frac{17}{4}$ (Put them on the same denominator.)

$R\left(\frac{1}{4}\right) = \frac{17}{4}$. This can also be expressed as $x = \frac{1}{4}$ and $y = \frac{17}{4}$ or $\left(\frac{1}{4}, \frac{17}{4}\right)$.

Example 11

Using the graph of the function $f(x)$ below, answer the following questions:

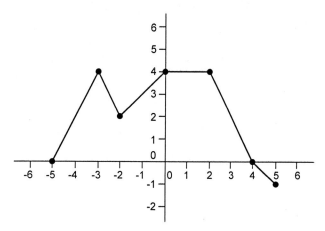

a) Find $f(2)$.

Finding $f(2)$ means find the y-value when x is 2. Based on the graph, $f(2)$ is 4.

b) Find $f(-2)$.

Finding $f(-2)$ means find the y-value when x is $-$ 2. Based on the graph, $f(-2)$ is 2.

c) Find the value(s) of x when $f(x)$ is -1.

In other words, find x when y is -1. Based on the graph, y is -1 when x is 5.

d) Find the value (s) of x when $f(x)$ is 0.

In other words, find x when y is 0. Based on the graph, y is 0 when x is 4 and -5.

Example 12

Ellen works in the learning center. Her salary can be modeled by the function $S(x) = 7.5x$, where $S(x)$ is the salary in dollars and x is the number of hours spent working in the center on a given day.

a) What is her salary if she works for 5 hours?
b) If her paycheck is $150 at the end of the week, how many hours did she work?

Solution

a) Replacing x with 5 in the given function, we get:

$$S(5) = 7.5(5)$$

$S(5) = \$37.5$

Her salary is \$37.50 if she works 5 hours.

b) This time we are given the salary and it is \$150 for the week. Replacing $S(x)$ with 150 and solving for x or number of hours worked, we get:

$S(x) = 7.5x$

$150 = 7.5x$

$\dfrac{150}{7.5} = x$ (Divide both sides by 7.5.)

$20 = x$

If her salary was \$150, she then worked 20 hours that week.

Example 13

Dexter recently graduated from Valencia College and has a student loan of \$5,000. He paid \$1,500 with his tax refund. He agreed to make a payment of \$250 every month in order to pay off the balance. His payment function is given by $P(x) = 250x + 1500$ where $P(x)$ is the payment in dollars and x is the number of months.

a) How much of the loan will he pay off after 7 months?
b) How long will it take him to pay off the entire loan of \$5,000?

Solution

a) Using the payment function and replacing x with 7, we get:
$P(x) = 250x + 1500$

$P(7) = 250(7) + 1500$

$P(7) = 1750 + 1500$

$P(7) = 3,250$

After 7 months, Dexter pays off \$3,250.

b) Replacing $P(x)$ with 5,000 and solving for x, the number of months, we get:
$P(x) = 250x + 1500.$

$5000 = 250x + 1500$ (Replace $P(x)$ with 5000.)

$5000 - 1500 = 250x$ (Subtract 1500 from both sides.)

$3500 = 250x$ (Combine like terms.)

$\dfrac{3500}{250} = x$ (Divide both sides by 250.)

$14 = x$

It will take Dexter 14 months to pay off the entire amount of the loan.

❖ *Finding the domain of functions*

1) <u>Polynomial functions</u>

We discussed earlier the domain of a relation and we mentioned that the domain refers to the x- values or first coordinate of the points. In this section, we discuss the domain of three types of functions. They are polynomial, rational, and radical functions.

A polynomial is an expression comprising of the sum of two or more terms, made of variables, constants, and nonnegative integer exponents. The domain of a polynomial function is all real numbers. It means that all possible x-values can be found on the function or work for the function. There are no restrictions on the domain for polynomial functions.

Examples of Polynomial Functions

$$L(x) = x - \frac{3}{7}$$

$$S(x) = x^2 + 4x - 5$$

$$E(x) = -\frac{1}{2}x^5 + x^4 - 7x^3 + x + \frac{2}{3}$$

Example 14

Find the domain of this function.

$$S(x) = x^2 + 4x - 5$$

This function is a polynomial function. **The domain consists of all real numbers** or $(-\infty, \infty)$.

Look at the graph of this function below, for every x there is a y- value. There is no break or gap on the graph of the function. In other words, for any value x that is plugged in the function above, there is a y-value.

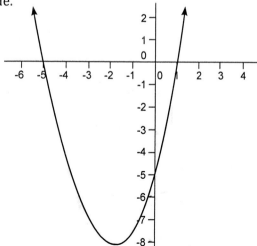

2) Rational functions

Rational functions are in the form of a fraction or ratio and have a variable in the denominator. As with any fraction, there cannot be a zero in the denominator ($\frac{5}{0}$ or a number times zero that results in 5. This is not possible thus undefined). Therefore, there cannot be a zero in the denominator of a rational function. To find the domain for this type of function, set the denominator equal to zero and solve it. The value(s) found, if any, will not be in the domain. In other words, the domain will be all real numbers except the value(s) found when solving the denominator of the function.

Examples of Rational Functions

$$M(x) = \frac{2x - 5}{x + 7}$$

$$F(x) = \frac{6}{x^2 - x - 42}$$

$$K(x) = \frac{x^3 + x}{x - 1}$$

Example 15

Find the domain of this function.
$$M(x) = \frac{2x - 5}{x + 2}$$

This function is a rational function. To find the domain, set the denominator of the function equal to zero and solve it.

$x + 2 = 0$

$x = -2$ 　　　(Subtract 2 from both sides.)

The value -2 if replaced for x in the function makes the denominator zero. The one x-value that does not work is -2. The domain of this function consists of **all real numbers except**
-2 or $\{x | x \neq -2\}$ or $(-\infty, -2) \cup (-2, \infty)$.

Look at the graph of this function below, for every x there is a y- value except for -2. The graph is actually split at -2. You can see how the graph stays away from the vertical line $x = -2$. This avoidance occurred because x cannot be -2. The line is a dotted line also called **vertical asymptote**, which stands for x-values that are not acceptable.

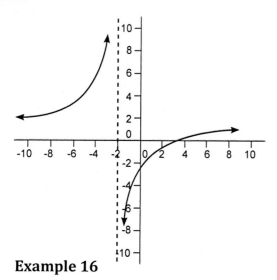

Example 16

Find the domain of this function.

$$F(x) = \frac{x - 5}{x - 3}$$

This function is a rational function. To find the domain, set the denominator of the function equal to zero and solve it.

$$x - 3 = 0$$

$$x = 3 \qquad \text{(Add 3 to both sides.)}$$

The value 3 if replaced for x in the function makes the denominator zero. The one x-value that does not work is 3. The domain of this function consists of **all real numbers except 3 or {x|x ≠ 3} or (-∞, 3) ∪ (3, ∞).**

Look at the graph of this function below, for every x there is y-value except for 3. The graph is actually split at 3. You can see how the graph stays away from the vertical line $x = 3$. This avoidance occurred because x cannot be 3. The line is a dotted line also called **vertical asymptote**, which stands for x-values that are not acceptable.

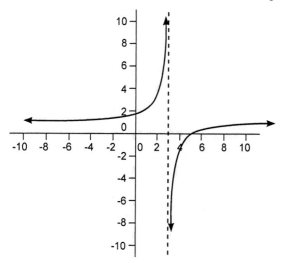

Example 17

Find the domain of this function.

$$S(x) = \frac{2x - 7}{3x + 4}$$

This function is a rational function. To find the domain, set the denominator of the function equal to zero and solve it.

$3x + 4 = 0$

$3x = -4$ (Subtract 4 from both sides.)

$x = -\frac{4}{3}$ (Divide both sides by 3.)

The value $-\frac{4}{3}$ if replaced for x in the function makes the denominator zero. The one x - value

that does not work is $-\frac{4}{3}$. The domain of this function consists of **all real numbers except** $-\frac{4}{3}$

or $\left\{x \middle| x \neq -\frac{4}{3}\right\}$ or $(-\infty, -\frac{4}{3}) \cup (-\frac{4}{3}, \infty)$.

Look at the graph of this function below, for every x there is y-value except for $-\frac{4}{3}$. The graph is

actually split at $-\frac{4}{3}$. You can see how the graph stays away from the vertical line $x = -\frac{4}{3}$. This

avoidance occurred because x cannot be $-\frac{4}{3}$. The line is a dotted line also called **vertical asymptote**, which stands for x-values that are not acceptable.

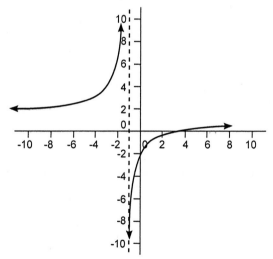

3) Radical functions

For these examples, we focus on the radical functions where the index is 2 (2 is not shown). The way we solve for domain holds true for all functions with **even indexes**. For functions with even index, there cannot be a negative number under the radical sign. An example is $\sqrt{-9}$. There is not a number multiplied by itself 2 times that results in -9 $[(3)(3) = 9; (-3)(-3) = 9)]$. To find the domain, **set the expression under the radical sign ≥ 0** (the expression **can** be positive or zero) and solve it.

We do not address domains of functions with odd indexes (3, 5, 7, etc.) and an example is $f(x) = \sqrt[3]{x + 6}$. However, keep in mind that the domain of a radical function with odd indexes is all real numbers or $(-\infty, \infty)$.

Examples of Radical Functions

$$D(x) = \sqrt{x + 10}$$

$$H(x) = \sqrt{6 - 7x}$$

$$N(x) = \sqrt{x} + \frac{2}{3}$$

Example 18

Find the domain of this function.

$$D(x) = \sqrt{x + 7}$$

There cannot be a negative number under the radical sign. To find the domain, set the expression under the radical sign ≥ 0 and solve it.

$x + 7 \geq 0$

$x \geq -7$ (Subtract 7 from both sides.)

The domain is **all x-values that are greater than or equal to -7 or $\{x | x \geq -7\}$ or $[-7, \infty)$.**

Look at the graph of the function below. From left to right, the graph starts on the x-axis at -7 and continues on.

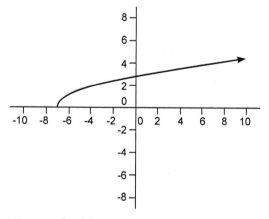

Example 19

Find the domain of this function.

$$H(x) = \sqrt{6 - x}$$

There cannot be a negative number under the radical sign. To find the domain, set the expression under the radical sign ≥ 0 and solve it.

$6 - x \geq 0$

$-x \geq -6$ (Subtract 6 from both sides.)

$-1(-x \geq -6)$ (Multiply or divide both sides by -1 and switch the inequality symbol.)

$x \leq 6$

The domain is **all x-values that are less than or equal to** 6 or $\{x | x \leq 6\}$ or $(-\infty, 6]$.

Look at the graph of the function below. From left to right, the graph does not go beyond 6 on the x-axis.

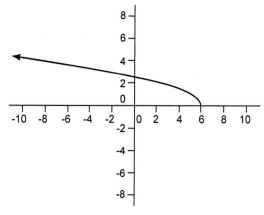

❖ *Graphing functions*

On way to graph a function is to use a table of values. We find points or ordered pairs, plot them, and connect the points. One example below is the graph of $f(x) = x^2$. Choose arbitrary x-values, replace them in the function to find y-values and use these points to draw the graph.

x	$f(x) = x^2$	(x, y)
-3	$f(-3) = 9$	$(-3, 9)$
-2	$f(-2) = 4$	$(-2, 4)$
-1	$f(-1) = 1$	$(-1, 1)$
0	$f(0) = 0$	$(0, 0)$
1	$f(1) = 1$	$(1, 1)$
2	$f(2) = 4$	$(2, 4)$
3	$f(3) = 9$	$(3, 9)$

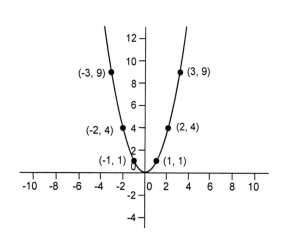

Another way to graph functions is to use a graphing calculator. In this course, a few functions are introduced. Be familiar with their names, their basic shapes, and how to graph them using a graphing calculator (*TI 83* or *TI 84*).

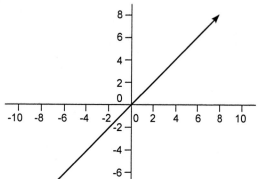

1. The **linear** function $L(x) = x$

- Press "y=".
- Press "X, T, θ, n" (next to Alpha for x).
- Press "graph".

2. The **square** or **quadratic** function $Q(x) = x^2$

- Press "y=".
- Press "X, T, θ, n" (next to Alpha for x).
- Press "x² ".
- Press "graph".

3. The **square root** function $S(x) = \sqrt{x}$

- Press "y=".
- Press "2nd"then"x² ".
- Press "X, T, θ, n" (next to Alpha for x).
- Press "graph".

4. The **cube root** function $C(x) = \sqrt[3]{x}$

- Press "y=",
- Press "MATH"then"4 ",
- Press "X, T, θ, n" (next to Alpha for x).
- Press "graph".

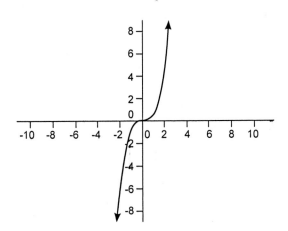

5. The **cube** function $C(x) = x^3$

- Press "y=".
- Press "X, T, θ, n" (next to Alpha for x).
- Press "^" (for raise).
- Press "3".
- Press "graph".

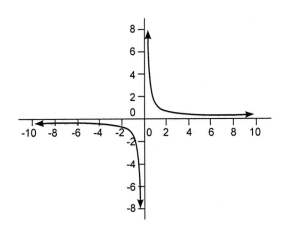

6. The **reciprocal** function $R(x) = \dfrac{1}{x}$

- Press "y=",
- Press "1" then "÷ ",
- Press "X, T, θ, n" (next to Alpha for x).
- Press "graph".

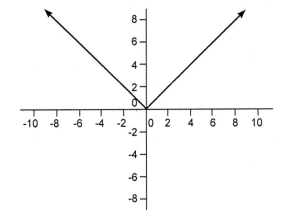

7. The **absolute value** function $V(x) = |x|$

- Press "y=".
- Press "2ND" then "0".
- Press "Enter".
- Press "X, T, θ, n" (next to Alpha for x).
- Press "graph".

Section 1.5 – Functions

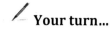 **Your turn...**

Is this relation a function?

1. $\{(-7, 9), (-6, 6), (-5, 5)\}$

2. $\{(0, 4), (1, 2), (3, 5), (4, -1), (3, -6), (2, -4)\}$

3. $\{(0, 5), (1, 6), (2, 7), (2, 11), (4, 13)\}$

4. Is this the graph of a function?

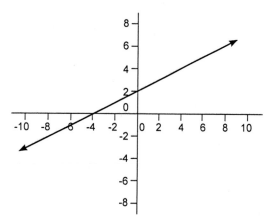

5. Is this the graph of a function?

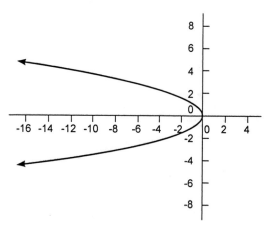

6. The function $W(x) = -7x + 3$ is given. Find $W(-1)$.

7. The function $P(x) = -7x^2 + 3x - 4$ is given. Find $P(-2)$.

8. The function $R(x) = 3x - 4$ is given. Find $R\left(\frac{1}{3}\right)$.

9. The function $R(x) = 3x - 4$ is given. Find $R\left(\frac{1}{7}\right)$.

10. Using the graph of the function $f(x)$ below, answer the following questions:

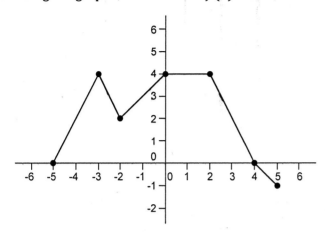

a) Find $f(0)$.

b) Find $f(-3)$.

c) Find the value(s) of x when $f(x)$ is 2.

11. Jenna works in the learning center. Her salary can be modeled by the function $S(x) = 9.5x$, where $S(x)$ is the salary in dollars and x is the number of hours spent working in the center on a given day.

a) What is her salary if she works for 6 hours?

b) If her paycheck is $104.50 at the end of the week, how many hours did she work?

12. Daniel recently graduated from Valencia College and has a student loan of $7,000. He paid $3,500 with his tax refund. He agreed to make a payment of $200 every month in order to pay off the balance. His payment function is given by $P(x) = 200x + 3500$ where $P(x)$ is the payment in dollars and x is the number of months.

a) How much of the loan will he pay off after 6 months?

b) How long will it take him to pay off the entire loan or $7,000?

Find the domain of the functions below.

13. $S(x) = x^2 + 2x - 5$

14. $M(x) = \dfrac{2x-5}{x+3}$

15. $F(x) = \dfrac{x-5}{x-2}$

16. $S(x) = \dfrac{2x-7}{4x+5}$

17. $D(x) = \sqrt{x+5}$

18. $H(x) = \sqrt{1-x}$

For more practice, access the online homework. See your syllabus for details.

Chapter 1: Review of Terms, Concepts, and Formulas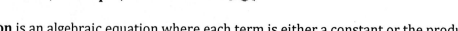

- A **linear equation** is an algebraic equation where each term is either a constant or the product of a constant and a variable raised to the first power. Linear equations can have one or more variables.

- Linear equations with one variable are in the form of $x = a$ or $y = b$. In these cases, $x = a$ is a **vertical line** and $y = b$ is a **horizontal line**.

- An **x-intercept** is the point at which the line crosses or touches the $x-$ axis, where the $y-$value equals 0.

- A **y-intercept** is the point at which the line crosses or touches the $y-$axis, where the $x-$ value equals 0.

- The **slope** of a line can be described as the incline or steepness of a line or how the graph changes.

- To find the **slope given a graph,** use $m = \dfrac{rise}{run}$ or $\dfrac{vertical\ distance}{horizontal\ distance}$.

- To find the **slope given 2 points,** use $m = \dfrac{y_2 - y_1}{x_2 - x_1}$.

- To find the **slope given an equation,** use the **slope-intercept form** $y = mx + b$. The slope is the coefficient of x and the y-intercept is $(0, b)$.

- **Parallel lines** have the same slope.

- **Perpendicular lines** have negative reciprocal slopes ($\frac{2}{3}$ and $-\frac{3}{2}$). In other words, multiply the two slopes and the result should be -1.

- To find an equation of the line, use the **point-slope form:**
  $y - y_1 = m(x - x_1)$, where (x_1, y_1) is the point and m is the slope.

- A **relation** can simply be defined as a set of points: $\{(x_1, y_1), (x_2, y_2), (x_3, y_3)\}$.

- The **domain** of a relation refers to all the x-values or the first coordinate of the points. To find the domain of a relation, simply list all the x-values.

- The **range** of a relation refers to all the y-values or the second coordinate of the points. To find the range of a relation, simply list all the y-values.

- A **function** has exactly one y-value for every x-value. A relation is said to be a function if every x-value has exactly one y-value. Therefore, a relation is not a function if one x-value has two or more corresponding y-values.

- If the relation is expressed graphically, use **the vertical line test** to determine if the graph is that of a function. A vertical line drawn anywhere on the graph should always intersect the graph at only one point. The graph is not a function if the line intersects the graph (not the x − axis) at two or more points.

- The **domain of a polynomial function** is all real numbers. It means that all possible x-values can be found on the function or work for the function.

- The **domain of a rational function** is all real numbers except the value(s) found when solving the denominator of the function.
  These values are also called vertical asymptotes. **Vertical asymptotes** stand for x-values that are not acceptable or that make the denominator zero.

- The **domain of a radical function** (index 2) is the result obtained after setting the expression under the radical sign ≥ 0 and solving it.

- For a **list of functions** covered in this chapter, see pages 56-57.

CHAPTER 2
SYSTEMS OF LINEAR EQUATIONS & INEQUALITIES

Section 2.1 - Systems of Linear Equations - Introduction

❖ Introducing systems of linear equations

❖ *Introducing systems of linear equations*

In chapter 1, we discussed linear equations and worked with one equation at a time. Well, in this chapter, we discuss systems of linear equations. A system is made of two or more equations and we will only address a system of 2 equations in this course. Examples of systems of linear equations are:

$$2x + 4y = 10$$
$$-5x + 7y = -3$$

$$3x + 6y = 12$$
$$-5x + 10y = 20$$

$$6x = 42$$
$$\frac{1}{7}x - y = -11$$

The focus of this chapter is to solve systems of linear equations. When solving a system of linear equations, look for the point of intersection or solution of both equations. In other words, find the point that is on both lines simultaneously.

Three methods can be used to solve systems of linear equations and they are **graphing, substitution, and elimination**. There are also **three possible kinds of solutions** and the next table shows a summary of the methods and solutions.

Methods and solutions when solving systems of linear equations graphically and algebraically

Possible Solutions	Solution Graphically	Solution using Substitution or Elimination
One solution	$y = -x + 2$ and $y = x - 6$ (4,-2)	$x = 4$ and $y = -2$ One specific point or point of intersection or $(4, -2)$
Infinitely many solutions	$-x + y = x + 3$ and $-2x + y = 3$	$0 = 0$ No specific x or y-value but a true statement The lines overlap or intersect each other at every point.
No solution	$-x + y = 2$ and $-x + y = -1$	$0 = -3$ No specific x or y-value but a false statement The lines do not intersect.

Example 1

Is the point (0, 2) a solution of this system of linear equations?

$2x + 5y = 10$

$-5x + 7y = -3$

Replace $x = 0$ and $y = 2$ in both equations. Starting with the first equation, we have:

$2x + 5y = 10$

$2(0) + 5(2) = 10$

$10 = 10$

The point is **a solution** for the first equation. Let us check the second equation. It needs to be a solution for both equations.

$-5x + 7y = -3$

$-5(0) + 7(2) = -3$

$14 \neq -3$

The point (0, 2) is **not a solution** for the second equation thus **not a solution** of the system of linear equations.

Example 2

Is the point (2, 1) a solution of this system of linear equations?

$3x + 6y = 12$

$5x + 10y = 20$

Replace $x = 2$ and $y = 1$ in both equations. Starting with the first equation, we have:

$3x + 6y = 12$

$3(2) + 6(1) = 12$

$6 + 6 = 12$

$12 = 12$

The point is a solution of the first equation. However, it needs to be the solution for both equations.

Let us check the second equation.

$5x + 10y = 20$

$5(2) + 10(1) = 20$

$10 + 10 = 20$

$20 = 20$

It is also a solution of the second equation. Since it is a **solution for both equations**, it is a **solution of the system of equations**.

Example 3

Is the point (7, 12) a solution of this system of linear equations?

$6x = 42$

$\dfrac{1}{7}x - y = -11$

Replace $x = 7$ and $y = 12$ in both equations where applicable. Starting with the first equation, we have:

$6(7) = 42$

$42 = 42$

The point is a solution of the first equation. However, it needs to be the solution for both equations. Let us check the second equation.

$\dfrac{1}{7}x - y = -11$

$\dfrac{1}{7}(7) - 12 = -11$

$1 - 12 = -11$

$-11 = -11$

It is also a solution of the second equation. Since it is a **solution for both equations**, it is a **solution of the system of equations**.

Section 2.1 - Systems of linear equations - introduction

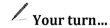 **Your turn...**

1. Is the point $(0, 4)$ a solution of this system of linear equations?

    $2x + 3y = 11$

    $-5x + 7y = -3$

2. Is the point $(2, 1)$ a solution of this system of linear equations?

    $x + 2y = 4$

    $5x + 10y = 20$

3. Is the point $(5, 12)$ a solution of this system of linear equations?

    $6x = 30$

    $\frac{1}{5}x - y = -11$

4. Is the point $(-5, 1)$ a solution of this system of linear equations?

    $6x = -30$

    $\frac{1}{5}x - y = -11$

5. Is the point $(5, 12)$ a solution of this system of linear equations?

    $6x + 2y = 3$

    $\frac{1}{3}x - y = -10$

For more practice, access the online homework. See your syllabus for details.

Section 2.2 - Systems of Linear Equations - Graphing

❖ Solving systems of linear equations by graphing

❖ *Solving systems of linear equations by graphing*

One method we use to solve a system of linear equations is the graphing method. First make sure that both equations are in the **slope-intercept form**. Once they are in this form, pay attention to the following:

a) Are the **equations parallel**; that is do they have the **same slope** but different y-intercepts? A good example is $y = -6x + 5$ and $y = -6x - 7$. In this case, graphing them would show two parallel lines. There is no need to go any further. Since the lines do not intersect, there is no point of intersection or solution. The answer is **no solution**.

b) Are the equations **identical**; that is, do they look exactly alike? A good example is $y = -\frac{5}{8}x + 7$ and $y = -\frac{5}{8}x + 7$. In this case, only one graph will be displayed when the graphs are input into the calculator. It means that they both overlap or intersect each other at every point. The answer is **infinitely many** solutions.

c) Do the equations have different slopes? One example is $y = 2x + 5$ and $y = \frac{1}{2}x - 1$. The equations can be graphed by hand. In this course, we will be using a graphing calculator (TI 83 or TI 84), so press the key "y=" and enter both equations into the calculator. Press the "graph" key. The lines **intersect at one point**. Therefore, there is one solution. To find the point of intersection or solution, follow these steps: press 2ND then Trace, press the number 5 key, and press the enter key 3 times. At the bottom of the screen the solution is displayed like this "*Intersection x = -4; y = -3*". The answer is $(-4, -3)$.

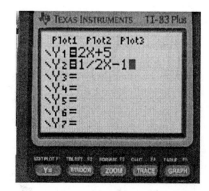 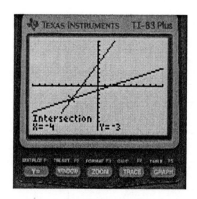

See chart on page 64 for a refresher on the 3 different kinds of possible solutions

Example 1

Solve by **graphing**.

$$7y - x = -63$$
$$y = \frac{1}{7}x + 2$$

The second equation is already in the slope-intercept form (y is isolated).

The first equation needs to be in the slope-intercept form.

$7y - x = -63$

$7y = x - 63$ (Add x to both sides.)

$y = \frac{x}{7} - \frac{63}{7}$ or $y = \frac{1}{7}x - 9$ (Divide both sides by 7.)

Enter $y = \frac{1}{7}x + 2$ and $y = \frac{1}{7}x - 9$ into the calculator. Press the "graph" key.

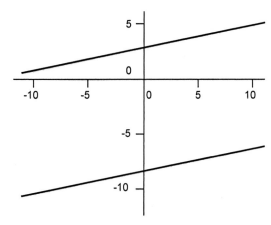

According to our graph, the lines are parallel because the slopes are the same. There is no point of intersection. The answer is then **no solution**.

Example 2

Solve by **graphing**.

$$-9x + y = 1$$
$$8y = 72x + 8$$

Both equations need to be in the slope-intercept form. Starting with the first equation, we have:

$-9x + y = 1$

$y = 9x + 1$ (Add $9x$ to both sides.)

Now isolating y in the second equation, we have:

$$8y = 72x + 8$$

$$y = \frac{72}{8}x + \frac{8}{8} \text{ or } y = 9x + 1 \qquad \text{(Divide both sides by 8.)}$$

Enter $y = 9x + 1$ and $y = 9x + 1$ into the calculator. Press the "graph" key.

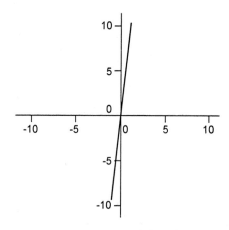

The lines are identical or they overlap. There is an **infinite number** of solutions.

Example 3

Solve by **graphing**.

$$y = -x + 5$$
$$y = x - 3$$

Both equations are in the slope-intercept form. Enter both equations into the calculator. Press the "graph" key. The lines intersect at a point.

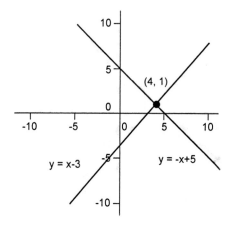

To find the point of intersection or solution, follow these steps: press 2ND then Trace, press the number 5 key, and press the enter key 3 times. At the bottom of the screen, the solution is displayed like this: *Intersection* $x = 4; y = 1$. The answer is $(4, 1)$.

Example 4

Solve by **graphing**.

$$3y = x + 3$$
$$y = x - 4$$

The second equation is in the slope-intercept form.

The first equation needs to be in the slope-intercept form. Isolating y in the first equation, we have:

$3y = x + 3$

$y = \frac{x}{3} + \frac{3}{3}$ or $y = \frac{1}{3}x + 1$ (Divide both sides by 3.)

Enter $y = x - 4$ and $y = \frac{1}{3}x + 1$ into the calculator. Press the "graph" key. The lines intersect at one point.

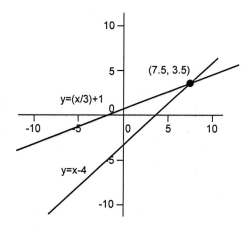

To find the point of intersection or solution, follow these steps: press 2ND then Trace, press the number 5 key, and press the enter key 3 times. At the bottom of the screen, the solution is displayed like this: *Intersection* $x = 7.5; y = 3.5$. The answer is $(7.5, 3.5)$.

Example 5

Solve by **graphing**.

$-x + y = 1$

$-6x + 3y = -21$

Both equations need to be in the slope-intercept form.

For the first equation, we have:

$-x + y = 1$

$y = x + 1$ (Add x to both sides.)

For the second equation, we have:

$-6x + 3y = -21$

$3y = 6x - 21$ (Add $6x$ to both sides.)

$y = \frac{6}{3}x - \frac{21}{3}$ or $y = 2x - 7$ (Divide both sides by 3.)

Enter $y = x + 1$ and $y = 2x - 7$ into the calculator. Press the "graph" key. The lines intersect at one point.

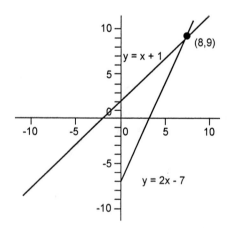

To find the point of intersection or solution, follow these steps: press 2ND then Trace, press the number 5 key, and press the enter key 3 times. At the bottom of the screen, the solution is displayed like this: *Intersection* $x = 8$; $y = 9$. The answer is $(8, 9)$.

Section 2.2 - Systems of linear equations - graphing

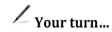 **Your turn...**

Solve by **graphing**. (See page 68 for calculator instructions)

System 1

$$3y - x = -9$$

$$y = \frac{1}{3}x + 2$$

System 2

$$-9x + y = 1$$

$$6y = 54x + 6$$

System 3

$$y = -x + 7$$

$$y = x - 2$$

System 4

$$2y = x + 2$$

$$y = x - 1$$

For more practice, access the online homework. See your syllabus for details.

Section 2.3 - Systems of Linear Equations - Elimination

❖ Solving systems of linear equations by elimination

❖ *Solving systems of linear equations by elimination*

Another method used to solve a system of linear equations is the elimination method also known as the **addition method**. First, write the equations in standard form, and then identify which variable to eliminate. The goal is to have the same coefficients with opposite signs in front of the variable chosen for elimination. Add the two linear equations, one variable will be eliminated and then solve for the other variable. Go back to the original equations, using either the first or the second equation, plug in the solved variable to get the other variable.

Recall that solving a system of linear equations means finding the **point of intersection (x, y).** If a specific x-value or y-value is not found, and instead there is a false statement like 0 = 5, the answer is **no solution**. The lines are parallel. If a specific x-value or y-value is not found, and instead there is a true statement like $0 = 0$ or $-8 = -8$, the answer is **infinitely many** solutions. The lines overlap or coincide.

See chart on page 64 for a refresher on the 3 different kinds of possible solutions.

Example 1

Solve using the **elimination method.**

$$7x + 3y = 4$$
$$\frac{7}{3}x + y = -5$$

We choose to eliminate y because it is easier. We only need to multiply the second equation by – 3 and add the 2 equations.

$$
\begin{array}{ll}
7x + 3y = 4 \quad \rightarrow \quad 7x + 3y = 4 & \text{(No changes were made to this equation.)} \\
\frac{7}{3}x + y = -5 \rightarrow \underline{-7x - 3y = 15} & \text{(Multiply by – 3 and add both equations.)} \\
\phantom{\frac{7}{3}x + y = -5 \rightarrow} 0 = 19 &
\end{array}
$$

Because **0 =19** is a false statement, there is **no solution** or point of intersection.

Example 2

Solve using the **elimination method.**

$$2x = -5y + 1$$
$$-10y = 4x - 2$$

Write both equations in standard form first ($ax + by = c$).

$$2x + 5y = 1 \qquad \text{(Add } 5y \text{ to both sides.)}$$
$$-4x - 10y = -2 \qquad \text{(Subtract } 4x \text{ from both sides.)}$$

In this case, we could eliminate x or y. It does not make a difference which variable is eliminated. In some cases, one variable is easier to eliminate than the other.

We choose to eliminate x, so multiply the first equation by 2 and add the 2 equations.

$$2x + 5y = 1 \quad \rightarrow \quad 4x + 10y = 2 \qquad \text{(Multiply both sides by 2.)}$$
$$-4x - 10y = -2 \rightarrow \underline{-4x - 10y = -2} \qquad \text{(No changes were made to this equation.)}$$
$$0 = 0$$

This is a true statement. There is no specific x and y-value. The answer is then **infinitely many solutions.**

Example 3

Solve using the **elimination method.**

$$-5x + 3y = -12$$
$$x + 7y = 10$$

We choose to eliminate x, so multiply the second equation by 5 and add the 2 equations.

$$-5x + 3y = -12 \quad \rightarrow \quad -5x + 3y = -12 \qquad \text{(No changes were made to this equation.)}$$
$$x + 7y = 10 \quad \rightarrow \quad \underline{5x + 35y = 50} \qquad \text{(Multiply both sides by 5.)}$$
$$38y = 38 \qquad \text{(Divide both sides by 38.)}$$
$$y = 1$$

We obtained a specific value for y. Now we need to solve for x. Use either the first or second equation to solve for x.

Choosing the second equation and replacing y with 1, we get:

$$x + 7y = 10$$

$$x + 7(1) = 10$$

$$x + 7 = 10$$

$$x = 10 - 7 \qquad \text{(Subtract 7 from both sides.)}$$

$$x = 3$$

There is **one point of intersection** or solution, and it is (**3, 1**).

Example 4

Solve using the **elimination method.**

$$-5x + 3y = 55$$
$$7x + y = 1$$

We choose to eliminate y because it is easier. We only need to multiply the second equation by – 3 and add the 2 equations.

$$-5x + 3y = 55 \;\rightarrow\; -5x + 3y = 55 \qquad \text{(No changes were made to this equation.)}$$

$$7x + y = 1 \quad \rightarrow\; \underline{-21x - 3y = -3} \qquad \text{(Multiply both sides by – 3.)}$$

$$-26x = 52 \qquad \text{(Divide both sides by } -26.)$$

$$x = -2$$

We obtained a specific value for x. Now we need to solve for y. Use either the first or second equation to solve for y.

Choosing the second equation and replacing x with -2, we get:

$$7x + y = 1$$

$$7(-2) + y = 1 \qquad \text{(Replace } x \text{ with } -2.)$$

$$-14 + y = 1$$

$$y = 1 + 14 \qquad \text{(Add 14 to both sides.)}$$

$$y = 15$$

There is **one point of intersection** or solution and it is (**-2, 15**).

Example 5

Solve using the **elimination method.**

$$2x + 3y = 9$$
$$5x + 4y = 5$$

This problem requires a little bit more work than the previous ones. There are many options.

- To cancel x, multiply the first equation by 5 and the second equation by -2 OR multiply the first equation by -5 and the second equation by 2.

- To cancel y, multiply the first equation by -4 and the second equation by 3 OR multiply the first equation by 4 and the second equation by -3.

We choose to multiply the first equation by 5 and the second equation by -2 to cancel x; then add the 2 equations.

$$2x + 3y = 9 \quad \rightarrow \quad 10x + 15y = 45 \qquad \text{(Multiply both sides by 5.)}$$
$$5x + 4y = 5 \rightarrow \quad \underline{-10x - 8y = -10} \qquad \text{(Multiply both sides by } -2.\text{)}$$
$$7y = 35 \qquad \text{(Divide both sides by 7.)}$$
$$y = 5$$

Now use either the first or second equation to solve for x. We use the first equation and we replace y with 5.

$$2x + 3y = 9$$

$$2x + 3(5) = 9$$

$$2x + 15 = 9$$

$$2x = 9 - 15 \qquad \text{(Subtract 15 from both sides.)}$$

$$2x = -6 \rightarrow x = -3 \qquad \text{(Divide both sides by 2.)}$$

There is **one point of intersection** or solution and it is $(-3, 5)$.

Example 6

The Rogue Scholars has a benefit concert on Osceola Campus. The admission fee for children is different from that of adults. A group with 2 children and 3 adults pays $19. A family with 4 children and 1 adult pays $13. Use a system of linear equations to find the price of admission for a child and for an adult.

Solution

Using x for the price of a child and y for the price of an adult, we set up the system of linear equations this way:

$$2x + 3y = 19$$
$$4x + y = 13$$

We can eliminate x or y. We choose to eliminate y because it is easier. We only need to multiply the second equation by -3, and then we add the 2 equations.

$$
\begin{array}{ll}
2x + 3y = 19 \rightarrow \quad 2x + 3y = 19 & \text{(No changes were made to this equation.)} \\
4x + y = 13 \rightarrow \underline{-12x - 3y = -39} & \text{(Multiply both sides by } -3.) \\
\qquad\qquad\qquad -10x = -20 & \text{(Divide both sides by } -10.)
\end{array}
$$

$$x = 2$$

Now plug in $x = 2$ in either the first or second equation. Using the first equation, we get:

$$2(2) + 3y = 19$$

$$4 + 3y = 19$$

$$3y = 19 - 4 \qquad\qquad \text{(Subtract 4 from both sides.)}$$

$$3y = 15 \rightarrow y = 5 \qquad\qquad \text{(Divide both sides by 3.)}$$

The price of admission for a **child** is **$2.00** and for an **adult $5.00**.

Example 7

Ryan took an algebra test and quiz last week. He answered 10 questions correctly on the test and 1 question correctly on the quiz. He earned a total of 52 points. Summer took the same test and quiz. She answered 12 questions on the test and 5 questions on the quiz correctly. She earned 70 points. How much was each question worth on the test? How much was each question worth on the quiz?

Using x for a test question and y for a quiz question, we set up the system of linear equations this way:

$$10x + y = 52$$
$$12x + 5y = 70$$

We choose to eliminate y because it is easier. We only need to multiply the first equation by -5, and then add the 2 equations.

$$10x + y = 52 \;\rightarrow\; -50x - 5y = -260 \qquad \text{(Multiply both sides by } -5.)$$
$$12x + 5y = 70 \;\rightarrow\; \underline{\;\; 12x + 5y = \quad\; 70 \;\;} \qquad \text{(No changes were made to this equation.)}$$
$$-38x = -190 \qquad \text{(Divide both sides by } -38.)$$
$$x = 5$$

Each **test question** was worth **5 points**.

Now plug in $x = 5$ in either the first or second equation.

Replacing x with 5 in the first equation, we get:

$$12(5) + 5y = 70$$

$$60 + 5y = 70$$

$$5y = 70 - 60 \qquad \text{(Subtract 60 from both sides.)}$$

$$5y = 10 \rightarrow y = 2 \qquad \text{(Divide both sides by 5.)}$$

Each **quiz question** was worth **2 points**.

Example 8

For the Spring Dance Concert, the Performing Arts Center (PAC) on East campus sold 200 tickets to Valencia faculty and students, and collected $2,230 in revenue. If the tickets sold to faculty cost $12 each and the ones sold to students cost $10 each, how many tickets were sold to faculty and to students?

Solution

Using x for the number of tickets sold to faculty and y for the number of tickets sold to students, we set up the system of linear equations this way:

$$x + y = 200$$
$$12x + 10y = 2230$$

We can eliminate x or y. We choose to eliminate y. We only need to multiply the first equation by -10, and then we add the 2 equations.

$$x + y = 200 \;\rightarrow\; -10x - 10y = -2000 \qquad \text{(Multiply both sides by } -10.)$$
$$12x + 10y = 2230 \;\rightarrow\; \underline{\;\; 12x + 10y = \;\; 2230 \;\;} \qquad \text{(No changes were made to this equation.)}$$
$$2x = 230 \qquad \text{(Divide both sides by 2.)}$$

$$x = 115$$

Now plug in $x = 115$ in either the first or second equation. Using the first equation, we get:

$115 + y = 200$

$y = 200 - 115$ (Subtract 115 from both sides.)

$y = 85$ (Combine like terms.)

The number of tickets sold to **faculty** is **115** and the number of tickets sold to **students** is **85**.

Section 2.3 - Systems of linear equations - elimination

 **Your turn...**

Solve using the **elimination method.**

System 1

$$5x + 3y = 1$$

$$\frac{5}{3}x + y = -7$$

System 2

$$x = -4y + 3$$

$$-8y = 2x - 6$$

System 3

$$-2x + 3y = 13$$

$$x + y = 6$$

System 4

$$2x + 5y = 1$$

$$3x + 4y = 12$$

System 5

The Rogue Scholars has a benefit concert on Osceola Campus. The admission fee for children is different from that of adults. A group with 4 children and 2 adults pays $22. A family with 3 children and 1 adult pays $14. Use a system of linear equations to find the price of admission for a child and for an adult.

System 6

Hank took an algebra test and quiz last week. He answered 15 questions correctly on the test and 3 questions correctly on the quiz. He earned a total of 63 points. Celine took the same test and quiz. She answered 14 questions on the test and 5 questions on the quiz correctly. She earned 61 points. How much was each question worth on the test? How much was each question worth on the quiz?

System 7

For the Spring Dance Concert, the Performing Arts Center (PAC) on East campus sold 250 tickets to Valencia faculty and students, and collected $1,610 in revenue. If the tickets sold to faculty cost $8 each and the ones sold to students cost $5 each, how many tickets were sold to faculty and students?

For more practice, access the online homework. See your syllabus for details.

Section 2.4 - Systems of Linear Equations - Substitution

❖ Solving systems of linear equations by substitution

❖ *Solving systems of linear equations by substitution*

Yet another method used to solve a system of linear equations is the substitution method. First, isolate one of the variables in one of the equations. Choose the variable that has a coefficient of 1 because it is easier to work with when using this method. Once a variable is isolated, substitute it in the other equation and solve. For instance, if y is isolated in the first equation, substitute it in the second equation. If x is isolated in the second equation, substitute it in the first equation. Once one of the variables is solved for, go back to the original equations, and use either equation to solve for the other variable.

Recall that solving a system of linear equations means finding the **point of intersection (x, y).** If a specific x-value or y-value is not found, and instead there is a false statement like $0 = 5$, the answer is **no solution**. The lines are parallel. If a specific x-value or y-value is not found, and instead there is a true statement like $0 = 0$ or $-8 = -8$, the answer is **infinitely many** solutions. The lines overlap or coincide.

See chart on page 64 for a refresher on the 3 different kinds of possible solutions.

Example 1

Solve using the **substitution method.**

$$7x + 3y = 4$$
$$\frac{7}{3}x + y = -5$$

Isolate y in the second equation. It is easier to work with because the coefficient is 1.

$$\frac{7}{3}x + y = -5$$

$$y = -5 - \frac{7}{3}x \qquad \text{(Subtract } \frac{7}{3}x \text{ from both sides.)}$$

Substitute $y = -5 - \frac{7}{3}x$ in the first equation. Now we are only dealing with one variable.

$$7x + 3y = 4$$

$$7x + 3(-5 - \frac{7}{3}x) = 4$$

$$7x - 15 - 7x = 4 \qquad \text{(Distribute the 3 and combine like terms.)}$$

$$-15 = 4$$

This is a false statement, so there is **no solution** or point of intersection.

Example 2

Solve using the **substitution method.**

$$-9x + y = 1$$
$$8y = 72x + 8$$

Isolate y in the second equation by dividing both sides by 8. Also, the variable y can be isolated in the first equation. Either option is fine.

$$8y = 72x + 8$$

$$y = \frac{72}{8}x + \frac{8}{8} \rightarrow y = 9x + 1 \qquad \text{(Divide both sides by 8.)}$$

Substitute $y = 9x + 1$ in the first equation. Now we are dealing with one variable.

$$-9x + y = 1$$

$$-9x + (\mathbf{9x + 1}) = 1 \qquad \text{(Combine like terms.)}$$

$$1 = 1$$

This is a true statement. There is no specific x and y-value. The answer is then **infinitely many solutions**.

Example 3

Solve using the **substitution method.**

$$-5x + 3y = -12$$
$$x + 7y = 10$$

Isolate x in the second equation. It is easier to work with because the coefficient 1.

$$x + 7y = 10$$

$$x = -7y + 10 \qquad \text{(Subtract } 7y \text{ from both sides.)}$$

Substitute $x = -7y + 10$ in the first equation. Now we are dealing with one variable.

$$-5x + 3y = -12$$

$$-5(\mathbf{-7y + 10}) + 3y = -12$$

$$35y - 50 + 3y = -12 \qquad \text{(Distribute the } -5.)$$

$$38y - 50 = -12 \qquad \text{(Combine like terms.)}$$

$$38y = -12 + 50 \qquad \text{(Add 50 to both sides.)}$$

$$38y = 38 \rightarrow y = 1 \qquad \text{(Replace this in the first or second equation.)}$$

Using the second equation, we get:

$x + 7y = 10$

$x + 7(1) = 10$

$x + 7 = 10$

$x = 10 - 7 \rightarrow x = 3$ (Subtract 7 from both sides.)

There is **one point of intersection** or solution and it is **(3, 1)**.

Example 4

Solve using the **substitution method.**

$$-5x + 3y = 55$$
$$7x + y = 1$$

Isolate y in the second equation because the coefficient is 1.

$7x + y = 1$

$y = -7x + 1$ (Subtract $7x$ from both sides.)

Substitute $y = -7x + 1$ in the first equation. Now we are dealing with one variable.

$-5x + 3(\mathbf{-7x + 1}) = 55$

$-5x - 21x + 3 = 55$ (Distribute the 3.)

$-26x + 3 = 55$ (Combine like terms.)

$-26x = 55 - 3$ (Subtract 3 from both sides.)

$-26x = 52 \rightarrow x = \mathbf{-2}$ (Divide both sides by -26.)

Replace $x = -2$ in the second equation. The first equation can also be used.

$7(-2) + y = 1$

$-14 + y = 1$

$y = 1 + 14$ (Add 14 to both sides.)

$y = \mathbf{15}$

There is **one point of intersection** or solution, and it is **(−2, 15)**.

Example 5

Solve using the **substitution method.**

$$2x + 3y = 9$$
$$5x + 4y = 5$$

Normally the elimination method is preferable when solving this problem, as already shown in *example 5 section 2.3* (Elimination method section). However, it is also good to know how to use the substitution method on a system where none of the variables has coefficient 1.

Isolate x in the first equation. It does not matter which variable is chosen in this case.

$2x + 3y = 9$

$2x = -3y + 9$ (Subtract $3y$ from both sides.)

$x = -\dfrac{3}{2}y + \dfrac{9}{2}$ (Divide both sides by 2.)

Substitute $x = -\dfrac{3}{2}y + \dfrac{9}{2}$ in the second equation. Now we are dealing with one variable.

$5(-\dfrac{3}{2}y + \dfrac{9}{2}) + 4y = 5$

$-\dfrac{15}{2}y + \dfrac{45}{2} + 4y = 5$ (Distribute the 5.)

$-15y + 45 + 8y = 10$ (Multiply both sides by 2 to clear the fractions.)

$-7y + 45 = 10$ (Combine like terms.)

$-7y = -45 + 10$ (Subtract 45 from both sides.)

$y = 5$ (Combine like terms and divide by -7.)

Now substitute $y = 5$ in either the first or second equation.

Using the first equation, we get:

$2x + 3y = 9$

$2x + 3(5) = 9$

$2x + 15 = 9$ (Distribute the 3.)

$2x = -15 + 9$ (Add 15 to both sides.)

$x = -3$ (Combine like terms and divide both sides by 2.)

There is **one point of intersection** or solution and it is $(-3, 5)$.

Example 6

The sum of two numbers is 33. One number is 5 more than 3 times the other number. What are the numbers?

We set up the 2 equations this way:

$$x + y = 33$$

$$x = 3y + 5$$

We use the **substitution method** since one of the variables is already isolated. Take x from the second equation and substitute it in the first equation.

$x + y = 33$	(First equation)
$(\mathbf{3y + 5}) + y = 33$	(Now we are dealing with one variable.)
$4y + 5 = 33$	(Combine like terms.)
$4y = 33 - 5$	(Subtract 5 from both sides of the equation.)
$4y = 28 \rightarrow \mathbf{y = 7}$	(Divide both sides of the equation by 4.)

To get the other number or x, use either the first or second equation. Using the second equation:

$$x = 3(7) + 5 \rightarrow \mathbf{x = 26}$$

The numbers are **7** and **26**.

Example 7

It takes 8 hours for Michel and Blake to paint a few rooms in building four of the Osceola campus. If it takes Michel 2 hours more than Blake to paint the rooms, how many hours did it take each one to paint the rooms?

We set up the 2 equations this way:

$$x + y = 8$$

$$x = y + 2 \qquad \text{(Blake is } y \text{ and Michel is } x\text{)}$$

We use the **substitution method** since one of the variables is already isolated. Take x from the second equation and substitute it into the first equation.

$x + y = 8$	(First equation)
$(\mathbf{y + 2}) + y = 8$	(Now we are dealing with one variable.)
$2y + 2 = 8$	(Combine like terms.)
$2y = 8 - 2$	(Subtract 2 from both sides.)

$2y = 6 \rightarrow y = 3$ (Divide both sides by 2.)

To get the other number or x, use either the first or second equation. Using the second equation:

$x = 3 + 2 \rightarrow x = 5$

Blake takes **3 hours** to paint the rooms while **Michel** takes **5 hours** to paint the rooms.

Example 8

The amount of $15,000 is invested in two funds paying 2% and 5% simple interest. If the annual interest is $540, how much of the $15,000 is invested at each interest rate?

Solution

Solving using the **substitution method**

Let x be the amount invested at 2% and y be the amount invested at 5%.

$$x + y = 15000$$

$$0.02x + 0.05y = 540$$

You can isolate either x or y in the first equation because they both have coefficients 1. We choose to isolate y in the first equation.

$x + y = 15000$

$y = -x + 15000$ (Subtract x from both sides.)

Substitute $y = -x + 15000$ in the second equation. Now we are dealing with one variable.

$0.02x + 0.05(-x + \mathbf{15000}) = 540$

$0.02x - 0.05x + 750 = 540$ (Distribute the 0.05.)

$-0.03x + 750 = 540$ (Combine like terms.)

$-0.03x = 540 - 750$ (Subtract 750 from both sides.)

$-0.03x = -210 \rightarrow x = \mathbf{7,000}$ (Divide both sides by -0.03.)

Replace $x = 7000$ in the first equation. The second equation can also be used.

$7000 + y = 15000$

$y = 15000 - 7000$ (Subtract 7000 from both sides.)

$y = \mathbf{8,000}$

The amount invested at **2%** is **$7,000** and the amount invested at **5%** is **$8,000**.

Section 2.4 - Systems of linear equations - substitution

 **Your turn...**

Solve using the **substitution method.**

System 1

$$5x + 3y = 1$$

$$\frac{5}{3}x + y = -7$$

System 2

$$x = -4y + 3$$

$$-8y = 2x - 6$$

System 3

$$-2x + 3y = 13$$

$$x + y = 6$$

System 4

$$2x - 3y = 1$$

$$-4x + 6y = -2$$

System 5

The sum of two numbers is 25. One number is 5 less than 2 times the other number. What are the numbers?

System 6

It takes 9 hours for Pascal and Regis to paint a few rooms in building four of the Osceola Campus. If it takes Pascal 3 hours more than Regis to paint the rooms, how many hours did it take each one to paint the rooms?

System 7

The amount of $13,000 is invested in two funds paying 3% and 6% simple interest. If the annual interest is $660, how much of the $13,000 is invested at each interest rate?

For more practice, access the online homework. See your syllabus for details.

Section 2.5 – Graphing Linear Inequalities and Systems of Linear Inequalities

❖ Graphing linear inequalities
❖ Graphing systems of linear inequalities

In this section, we discuss how to graph the solution set for linear inequalities and for a system of linear inequalities. Solving in this case means finding all the points that make the inequality true. For instance, the points $(0, 1)$ and $(-3, -5)$ are solutions of the linear inequality $y < 3x + 7$. When x and y are replaced by their respective values into the inequality, we have a true statement (such as $1 < 7$). We will be using a graphing calculator for this. (To Graph by hand, See Appendix B)

To show the solutions, graph the inequality and shade the area that has all the solutions. To begin, make sure that y is isolated in the inequality. Enter it into the graphing calculator (TI 83 or 84). If the symbol is $<$ or $>$, the line is a dotted line because there is no equal sign. If the symbol is $\leq$ or $\geq$, then the line is a solid line because of the equal sign.

The symbols $<$ or $\leq$ mean less than so shade below the line. This feature is found by moving the left cursor in front of y_1 and then press the "enter" key three times (Small triangle lower left). The symbols $>$ or $\geq$ mean greater than so shade above the line. This symbol is found by moving the left cursor in front of y_1 and then press the "enter" key two times (Small triangle upper right).

Examples of graphs of inequalities using a graphing calculator are shown below.

$$y \leq 2x + 5 \quad \text{(Shade below)}$$

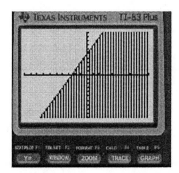

$$y \geq 2x + 5 \quad \text{(Shade above)}$$

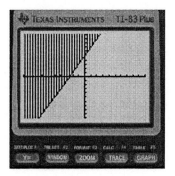

For a system of linear inequalities, make sure that y is isolated in both inequalities. Enter both inequalities into the calculator. Use the above steps for each individual inequality. After shading, the solutions that are common to both inequalities are in the darkest or overlapping area.

An example of a graph of a system of linear inequalities using a graphing calculator is show below.

$$y \geq x + 5$$

$$y \leq -x + 3$$

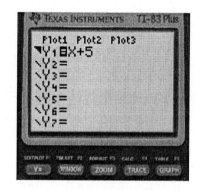
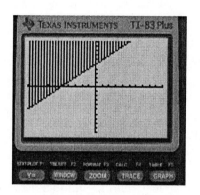

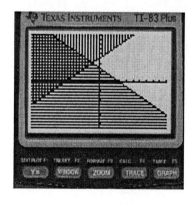

❖ *Graphing linear inequalities*

Example 1

Graph the linear inequality $y > 4x + 4$.   *(See page 89 for calculator instructions)*

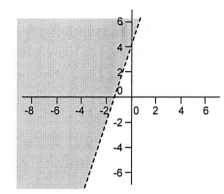

Check a couple of points to verify the accuracy of this graph. Test the point $(0, 0)$.

$$0 > 4(0) + 4 \quad \rightarrow \quad 0 > 4$$

This is a false statement so $(0, 0)$ is not in the shaded area of the graph.

Test the point $(-2, 2)$.

$$2 > 4(-2) + 4 \quad \rightarrow \quad 2 > -4$$

This is a true statement so $(-2, 2)$ is in the shaded area of the graph.

Example 2

Graph the linear inequality $2x + y \leq 3$.   *(See page 89 for calculator instructions)*

Put the inequality in the slope-intercept form.

$$y \leq -2x + 3 \qquad \qquad \text{(Subtract } 2x \text{ from both sides.)}$$

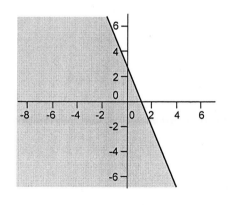

Check a couple of points to verify the accuracy of this graph. Test the point $(0, 0)$.

$$0 \leq -2(0) + 3 \quad \rightarrow \quad 0 \leq 3$$

This is true so the point $(0,0)$ is in the shaded area of the graph.

Test the point $(2, 2)$.

$$2 \leq -2\,(2) + 3 \rightarrow 2 \leq -1$$

This is false so the point $(2,2)$ is not in the shaded area of the graph.

Example 3

Graph the linear inequality $x - 3y < 6$.   *(See page 89 for calculator instructions)*

Put the inequality in the slope-intercept form.

$x - 3y < 6$

$-3y < -x + 6$ (Subtract x from both sides.)

$y > -\dfrac{x}{-3} + \dfrac{6}{-3}$ (Divide both sides by -3.)

$y > \dfrac{1}{3}x - 2$ (Switch the inequality symbol when dividing by a negative number.)

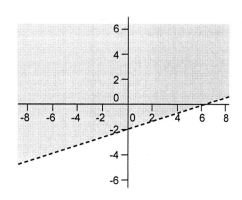

Check a couple of points to verify the accuracy of this graph. Test the point $(0, 0)$.

$0 > \dfrac{1}{3}(0) - 2 \;\rightarrow\; 0 > -2$

This is true so the point $(0,0)$ is in the shaded area of the graph.

Test the point $(0, -4)$.

$-4 > \dfrac{1}{3}(0) - 2 \;\rightarrow\; -4 > -2$

This is false so the point $(0, -4)$ is not in the shaded area of the graph.

❖ *Graphing systems of linear inequalities*

Example 4

Graph the system of linear inequalities.   *(See pages 89-90 for calculator instructions)*

$\qquad y \leq x + 3$

$\qquad y \geq -x + 5$

They are already in slope-intercept form. Enter both inequalities in the calculator then graph.

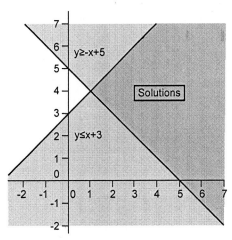

Check a point to verify the accuracy of this graph. Test the point (4, 4) for both inequalities.

$$y \leq x + 3 \quad \rightarrow \quad 4 \leq 4 + 3 \quad \rightarrow \quad 4 \leq 7$$

This is true for the first inequality.

$$y \geq -x + 5 \quad \rightarrow 4 \geq -4 + 5 \quad \rightarrow \quad 4 \geq 1$$

This is also true for the second inequality.

This is true for both inequalities, so the point (4, 4) is in the darkest shaded area of the graph.

Example 5

Graph the system of linear inequalities. *(See pages 89-90 for calculator instructions)*

$$2x + y > 7$$
$$3x - 3y < 12$$

Isolate y in both equations so they can be in the slope-intercept form.

$2x + y > 7 \rightarrow \boldsymbol{y > -2x + 7}$ (Subtract $2x$ from both sides.)

$3x - 3y < 12$

$-3y < -3x + 12$ (Subtract $3x$ from both sides.)

$y > -\dfrac{3}{-3}x + \dfrac{12}{-3} \rightarrow \boldsymbol{y > x - 4}$ (Divide both sides by -3 and switch the inequality symbol.)

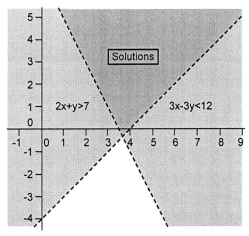

Enter $\boldsymbol{y > -2x + 7}$ and $\boldsymbol{y > x - 4}$ into the calculator then graph.

Check a point to verify the accuracy of this graph. Test the point (4, 2) for both inequalities.

$$y > -2x + 7 \quad \rightarrow \quad 2 > -2(4) + 7 \rightarrow 2 > -1$$

This is true for the first inequality.

$$y > x - 4 \quad \rightarrow \quad 2 > 4 - 4 \quad \rightarrow 2 > 0$$

This is also true for the second inequality.

This is true for both inequalities, so the point (4, 2) is in the darkest shaded area of the graph.

Example 6

Graph the system of linear inequalities.   *(See pages 89-90 for calculator instructions)*

$$-7x + y \geq 9$$
$$y \leq 7x - 5$$

The second equation is already in the slope-intercept from. Isolate y in the first equation.

$$-7x + y \geq 9$$

$$y \geq 7x + 9 \qquad \text{(Add } 7x \text{ to both sides.)}$$

Enter $y \leq 7x - 5$ and $y \geq 7x + 9$ into the calculator then graph. We obtain the graph below.

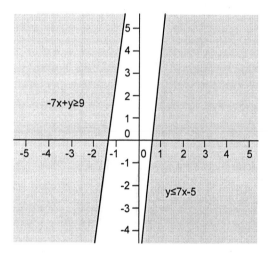

Since there is no overlap, there is **no solution**.

Section 2.5 – Graphing linear inequalities and systems of linear inequalities

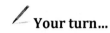 **Your turn...**

Graph the linear inequalities. (See pages 89-90 for calculator instructions)

1. $y > 2x + 8$
2. $3x + y \leq 5$
3. $x - 7y < 14$

Graph the systems of linear inequalities. (See pages 89-90 for calculator instructions)

System 1

$y \leq x + 1$

$y \geq -x + 7$

System 2

$2x + y > 6$

$5x - 5y < 15$

System 3

$-3x + y \geq 9$

$y \leq 3x - 2$

For more practice, access the online homework. See your syllabus for details.

Chapter 2: Review of Terms, Concepts, and Formulas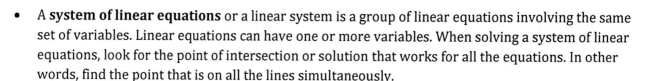

- A **system of linear equations** or a linear system is a group of linear equations involving the same set of variables. Linear equations can have one or more variables. When solving a system of linear equations, look for the point of intersection or solution that works for all the equations. In other words, find the point that is on all the lines simultaneously.

- A few of the methods that can be used to solve linear systems are: **graphing, substitution**, and **elimination**.

- To solve a system of linear equations **by graphing** using a TI 83 or 84 graphing calculator, take the following steps. Make sure that both equations are in the slope-intercept form. Then press the key "y=" and enter both equations into the calculator. Press the "graph" key. The lines **intersect at one point.** Therefore, there is one solution. To find the point of intersection or solution, follow these steps: press 2ND then Trace, press the number 5 key, and press the enter key 3 times.

- To solve a system of linear equations using the **elimination method**, first, if necessary put the equations in standard form ($ax + by = c$), and then identify which variable to eliminate. The goal is to have the same coefficients with opposite signs in front of the variable chosen for elimination. Add the two linear equations, one variable will be eliminated and then solve for the other variable. Go back to the original equations, using either the first or the second equation, plug in the solved variable to get the other variable.

- To solve a system of linear equations using the **substitution method**, first, isolate one of the variables in one of the equations. Choose the variable that has a coefficient of 1 because it is easier to work with when using this method. Once a variable is isolated, substitute it in the other equation and solve. For instance, if y is isolated in the first equation, substitute it in the second equation. If x is isolated in the second equation, substitute it in the first equation. Once one of the variables is solved for, go back to the original equations, and use either equation to solve for the other variable.

- A **linear inequality** is a linear expression in two variables that uses the symbols such as $<, >, \leq$ or $\geq$.

- When **graphing a linear inequality**, make sure that y is isolated in the inequality. Enter it into the graphing calculator (TI 83 or 84). If the symbol is $<$ or $>$, the line is a dotted line because there is no equal sign. If the symbol is $\leq$ or $\geq$, then the line is a solid line because of the equal sign. The symbols $<$ or $\leq$ mean less than so shade below the line. This feature is found by moving the left cursor in front of y_1 and then press the "enter" key three times (Small triangle lower left). The symbols $>$ or $\geq$ mean greater than so shade above the line. This symbol is found by moving the left cursor in front of y_1 and then press the "enter" key two times (Small triangle upper right).

- When **graphing a system of linear inequalities**, make sure that y is isolated in both inequalities. Enter both inequalities into the calculator. Use the above steps for each individual inequality. After shading, the solutions that are common to both inequalities are in the darkest or overlapping area.

CHAPTER 3

POLYNOMIALS

A polynomial is an expression comprising of the sum of two or more terms, made of variables, constants, and nonnegative integer exponents. We will address polynomials with a single variable. An example of a polynomial is:

$$9x^8 + 8x^7 - x^2 + 2x - 7$$

❖ *Terms*

A term is the product of a coefficient and a variable or variables.

The terms of the above polynomials are: $9x^8, 8x^7, -x^2, 2x, -7$

❖ **Coefficients**

The coefficient is the number in front of the variable. Choosing a few terms of the above polynomial, let us identify the coefficients.

The coefficient of $9x^8$ is 9.

The coefficient of $-x^2$ is -1.

The coefficient of -7 is -7.   (This is the same as the coefficient of $-7x^0$ or $-7(1)$ or -7)

Example 1

The polynomial $-11x^3 + 12x^2 + \frac{1}{2}x$ is given. List the terms of this polynomial. Give the coefficients of all the terms.

The terms are $-11x^3, 12x^2$, and $\frac{1}{2}x$.

The coefficients of all the terms are $-11, 12$, and $\frac{1}{2}$.

❖ *Degree*

The degree of a term refers to the exponent or power of the term. The degree of the polynomial is the highest degree when comparing the degree of all the terms.

The degree of $5x^4$ is 4.

The degree of $2x$ is 1.

The degree of -7 is 0.   [Same as $-7x^0$ or $-7(1)$ or -7]

The degree of the entire polynomial $9x^8 + 8x^7 - x^2 + 2x - 7$ is 8.

❖ *Classification of polynomials (based on the number of terms)*

$9x^8 \rightarrow$ One term thus it is classified as a monomial.

$9x^8 + 8x^7 \rightarrow$ This is the sum of two terms, so it is a binomial.

$-x^2 + 2x - 7 \rightarrow$ This is the sum of three terms, so it is a trinomial.

$9x^8 + 8x^7 - x^2 + 2x - 7 \rightarrow$ This is the sum of four or more terms, so we will use the name polynomial.

Example 2

The polynomial $-11x^3 + 12x^2 + \frac{1}{2}x$ is given. What is the degree of this polynomial? Classify this polynomial.

The degree of the polynomial is 3.

Since this polynomial has three terms, it is a trinomial.

❖ *Descending order*

The polynomial is in descending order when the degree of the terms starts from the highest to the lowest.

$8x^2 + 5x^4 - 7x^5 + x^3 \rightarrow$ This is **not** in descending order.

$-7x^5 + 5x^4 + x^3 + 8x^2 \rightarrow$ This is in descending order.

❖ *Leading term & leading coefficient*

The leading term is the term with the highest degree. The leading coefficient is the number in front of the variable of the leading term or the number in front of the variable with the highest degree.

$9x^8 + 8x^7 - x^2 + 2x - 7$ → This polynomial has a leading term of $9x^8$ and a leading coefficient of 9.

Example 3

The polynomial $-\frac{5}{3} + 5x^4 + x^3 + \frac{1}{2}x^5$ is given. Write this polynomial in descending order. Give the leading term and leading coefficient.

In descending order, the polynomial is written this way: $\frac{1}{2}x^5 + 5x^4 + x^3 - \frac{5}{3}$

The leading term is $\frac{1}{2}x^5$ and the leading coefficient is $\frac{1}{2}$.

Section 3.1 – Introducing polynomials

Your turn...

1. The polynomial $-10x^3 + 18x^2 + \frac{1}{5}x$ is given. List the terms of this polynomial. Give the coefficient of the third term.

2. The polynomial $-10x^3 + 18x^2 + \frac{1}{5}x$ is given. What is the degree of this polynomial? Classify this polynomial.

3. The polynomial $-\frac{5}{7} + 6x^2 - 5x + 3x^4 + 2x^3 + \frac{1}{3}x^5$ is given. Write this polynomial in descending order. Give the leading term and leading coefficient.

4. The polynomial $-20x^3 + 8x^2 - 7x$ is given. List the terms of this polynomial. Give the coefficient of the third term.

5. The polynomial $9x^4 - 50x^3$ is given. What is the degree of this polynomial? Classify this polynomial.

6. The polynomial $-1 + 3x^2 - 11x + 13x^4 + 4x^3 + 17x^5$ is given. Write this polynomial in descending order. Give the leading term and leading coefficient.

For more practice, access the online homework. See your syllabus for details.

Section 3.2 - Adding and Subtracting Polynomials

- ❖ Adding polynomials
- ❖ Subtracting polynomials

In order to add or subtract polynomials, the polynomial must contain like terms. In other words, the terms must have the same variables and degrees. If they are like terms, add the like terms by adding the coefficients. The variable never changes in addition or subtraction, but the coefficients change. Examples of like terms are:

$$3x^4, -5x^4$$

$$\frac{1}{2}t^2, 7t^2$$

❖ *Adding polynomials*

Example 1

Add the polynomials.

$(-4y^7 - y^3 + 7y^2) + (2y^7 + 3y^3 + 10y^2)$

$-4y^7 - y^3 + 7y^2 + 2y^7 + 3y^3 + 10y^2$ (Drop the parentheses.)

$-2y^7 + 2y^3 + 17y^2$ (Combine like terms.)

Example 2

Add the polynomials.

$(2m^5n^4 + 3m^3n^2) + (-7m^5n^4 - 11m^3n^2)$

$2m^5n^4 + 3m^3n^2 - 7m^5n^4 - 11m^3n^2$ (Drop the parentheses.)

$-5m^5n^4 - 8m^3n^2$ (Combine like terms.)

Example 3

Add the polynomials.

$$\left(\frac{1}{6}s^3 + \frac{1}{2}s + 7\right) + \left(\frac{1}{3}s^3 + \frac{5}{2}s + \frac{2}{5}\right)$$

$$\frac{1}{6}s^3 + \frac{1}{2}s + 7 + \frac{1}{3}s^3 + \frac{5}{2}s + \frac{2}{5} \qquad \text{(Drop the parentheses.)}$$

$$\frac{1}{2}s^3 + 3s + \frac{37}{5} \qquad \text{(Combine like terms.)}$$

❖ *Subtracting polynomials*

Example 4

Subtract the polynomials.

$$(-4y^7 - y^3 + 7y^2) - (2y^7 + 3y^3 + 10y^2)$$

$$-4y^7 - y^3 + 7y^2 - 2y^7 - 3y^3 - 10y^2 \qquad \text{(Distribute the negative 1.)}$$

$$-6y^7 - 4y^3 - 3y^2 \qquad \text{(Combine like terms.)}$$

Example 5

Subtract the polynomials.

$$(2m^5n^4 + 3m^3n^2) - (-7m^5n^4 - 11m^3n^2)$$

$$2m^5n^4 + 3m^3n^2 + 7m^5n^4 + 11m^3n^2 \qquad \text{(Distribute the negative 1.)}$$

$$9m^5n^4 + 14m^3n^2 \qquad \text{(Combine like terms.)}$$

Example 6

Subtract the polynomials.

$$\left(\frac{1}{5}s^3 + \frac{1}{2}s + 7\right) - \left(\frac{1}{3}s^3 + \frac{5}{2}s + \frac{2}{3}\right)$$

$$\frac{1}{5}s^3 + \frac{1}{2}s + 7 - \frac{1}{3}s^3 - \frac{5}{2}s - \frac{2}{3} \qquad \text{(Distribute the negative 1.)}$$

$$-\frac{2}{15}s^3 - 2s + \frac{19}{3} \qquad \text{(Combine like terms.)}$$

Example 7

Find the **perimeter** of the rectangle below.

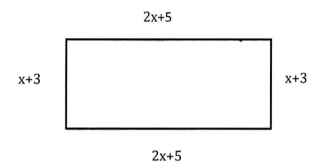

The formula for the perimeter of a rectangle is $P = 2l + 2w$ or $P = l + w + l + w$. In our example, we have $l = 2x + 5$ and $w = x + 3$.

$P = (2x + 5) + (x + 3) + (2x + 5) + (x + 3)$

$P = 2x + 5 + x + 3 + 2x + 5 + x + 3$ (Drop the parentheses.)

$P = 6x + 16$ (Combine like terms.)

Example 8

If x is the first consecutive integer, write the sum of four consecutive integers and simplify.

Solution

$x + (x + 1) + (x + 2) + (x + 3)$

$x + x + 1 + x + 2 + x + 3$ (Drop the parentheses.)

$4x + 6$ (Combine like terms.)

Section 3.2 – Adding and subtracting polynomials

 **Your turn...**

Add and subtract the polynomials.

1. $(-5y^7 + 2y^5 - 3y^3 + 9y^2) + (2y^7 + 5y^3 + 11y^2)$

2. $(3m^5n^4 + 4m^3n^2 + 5mn) + (-6m^5n^4 - 13m^3n^2)$

3. $\left(\frac{1}{6}s^3 - \frac{1}{3}s^2 + \frac{3}{2}s + 6\right) + \left(\frac{2}{3}s^3 + \frac{2}{3}s^2 + \frac{5}{2}s + \frac{2}{5}\right)$

4. $(-4y^7 + 7y^5 - 9y^3 + 7y^2) - (y^7 + 2y^3 + 15y^2)$

5. $(4m^5n^4 + 8m^3n^2 + 12mn) - (-11m^5n^4 - 3m^3n^2)$

6. $\left(\frac{1}{6}s^3 - \frac{1}{3}s^2 + \frac{3}{2}s + 6\right) - \left(\frac{2}{3}s^3 + \frac{2}{3}s^2 + \frac{5}{2}s + \frac{2}{5}\right)$

7. Find the **perimeter** of the rectangle below.

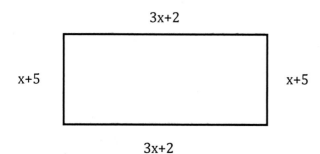

3x+2

x+5

x+5

3x+2

8. If x is the first consecutive even integer, write the sum of three consecutive even integers and simplify.

For more practice, access the online homework. See your syllabus for details.

Section 3.3 – Multiplying Polynomials

❖ Two monomials
❖ A monomial and a binomial
❖ A monomial and a trinomial
❖ Two binomials

To multiply polynomials, multiply the coefficients together and the variables together. If the terms have the same variable, keep the base or variable, and add the exponents. In the process of multiplying polynomials, we will also use the distributive property and the FOIL method when applicable.

We can also use the formulas below:

- Sum and difference: $(a + b)(a - b) = a^2 - b^2$

- Binomial squared: $(a + b)^2 = a^2 + 2ab + b^2$

- Binomial squared: $(a - b)^2 = a^2 - 2ab + b^2$

❖ *Two monomials*

Example 1

Multiply.

$(3x^2)(5yz^4)$

$15\ x^2yz^4$ (Multiply number with number and variable with variable.)

Example 2

Multiply.

$(-m^3n^7p^6)(m^2n^3p)$

$-m^5n^{10}p^7$ (Multiply by keeping the same base and adding the exponents.)

❖ *A monomial and a binomial*

Example 3

Multiply.

$a^{10}b^{11}c^{12}\left(abc - \frac{1}{2}b^7c^3\right)$

$(a^{10}b^{11}c^{12})\left(abc - \frac{1}{2}b^7c^3\right)$ (Distributive property)

$a^{11}b^{12}c^{13} - \frac{1}{2}a^{10}b^{18}c^{15}$ (Multiply by keeping the same base and adding the exponents.)

❖ *A monomial and a trinomial*

Example 4

Multiply.

$x^2y\left(-5x + 10y^3 - \frac{7}{3}\right)$

$-5x^3y + 10x^2y^4 - \frac{7}{3}x^2y$ (Distributive property)

❖ *Two binomials*

Example 5

Multiply.

$(x + 5)(x - 7)$

$(x)(x) + (x)(-7) + (5)(x) + (5)(-7)$ (FOIL – First, Outside, Inside, Last)

$x^2 - 7x + 5x - 35$ (Multiply.)

$x^2 - 2x - 35$ (Combine like terms.)

Example 6

Multiply.

$(3x - 2)(4x - 9)$

$(3x)(4x) + (3x)(-9) + (-2)(4x) + (-2)(-9)$ (FOIL – First, Outside, Inside, Last)

$12x^2 - 27x - 8x + 18$ (Multiply.)

$12x^2 - 35x + 18$ (Combine like terms.)

Example 7

Multiply.

$(y + 8)(y - 8)$

$(y)(y) + (y)(-8) + (8)(y) + (8)(-8)$ (FOIL – First, Outside, Inside, Last)

$y^2 - 8y + 8y - 64$ (Multiply.)

$y^2 - 64$ (Combine like terms.)

Example 8

Multiply.

$(2y - 6x)(2y + 6x)$

$(2y)(2y) + (2y)(6x) + (-6x)(2y) + (-6x)(6x)$ (FOIL – First, Outside, Inside, Last)

$4y^2 + 12xy - 12xy - 36x^2$ (Multiply.)

$4y^2 - 36x^2$ (Combine like terms.)

Example 9

Multiply.

$(4x - 7)^2$

$(4x - 7)(4x - 7)$ (Expand.)

$(4x)(4x) + (4x)(-7) + (-7)(4x) + (-7)(-7)$ (FOIL – First, Outside, Inside, Last)

$16x^2 - 28x - 28x + 49$ (Multiply.)

$16x^2 - 56x + 49$ (Combine like terms.)

Example 10

Multiply.

$\left(\frac{1}{2}x + 3\right)^2$

$\left(\frac{1}{2}x + 3\right)\left(\frac{1}{2}x + 3\right)$ (Expand.)

$\left(\frac{1}{2}x\right)\left(\frac{1}{2}x\right) + \left(\frac{1}{2}x\right)(3) + (3)\left(\frac{1}{2}x\right) + (3)(3)$ (FOIL – First, Outside, Inside, Last)

$\frac{1}{4}x^2 + \frac{3}{2}x + \frac{3}{2}x + 9$ (Multiply.)

$\frac{1}{4}x^2 + 3x + 9$ (Combine like terms.)

Example 11

Find the **area** of the rectangle below.

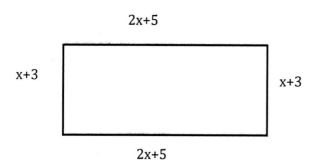

The formula for the area of a rectangle is $A = l \cdot w$. In our example, we have $l = 2x + 5$ and $w = x + 3$.

$A = (2x + 5)(x + 3)$

$A = 2x(x) + 2x(3) + 5(x) + 5(3)$ (FOIL – First, Outside, Inside, Last)

$A = 2x^2 + 6x + 5x + 15$ (Multiply.)

$A = 2x^2 + 11x + 15$ (Combine like terms.)

Example 12

Find the **area** of the square below.

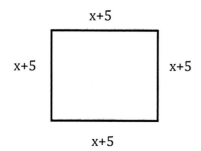

The formula for the area of a square is $A = s^2$. In our example, we have $s = x + 5$.

$A = (s)^2$

$A = (x + 5)^2$ (Replace s with $x + 5$.)

$A = (x + 5)(x + 5)$ (Expand.)

$A = x(x) + x(5) + 5(x) + 5(5)$ (FOIL – First, Outside, Inside, Last)

$A = x^2 + 5x + 5x + 25$ (Multiply.)

$A = x^2 + 10x + 25$ (Combine like terms.)

Example 13

The **perimeter** of the square below is 76 cm. Find x.

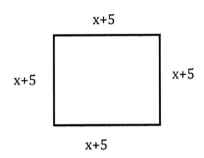

The formula for the perimeter of a square is $P = 4s$. In our example, we have $s = x + 5$. We are given the perimeter and it is 76 cm.

$P = 4s$

$76 = 4(x + 5)$ (Replace s with $x + 5$.)

$76 = 4x + 20$ (Distribute the 4.)

$76 - 20 = 4x$ (Subtract 20 from both sides.)

$56 = 4x$ (Combine like terms.)

$\frac{56}{4} = x$ (Divide both sides by 4.)

$14 = x$ or $x = \textbf{14 cm}$

Section 3.3 – Multiplying polynomials

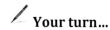 **Your turn...**

Multiply.

1. $(2x^2)(9yz^4)$

2. $(-m^3n^7p^5)(3m^2n^4p)$

3. $a^{11}b^{21}c^{15}(abc - \frac{1}{2}b^7c^5)$

4. $x^2y\left(-2x + 9y^3 - \frac{7}{4}\right)$

5. $(x + 3)(x - 5)$

6. $(5x - 1)(4x - 7)$

7. $(y + 6)(y - 6)$

8. $(3x - 4y)(3x + 4y)$

9. $(2x - 5)^2$

10. $\left(\frac{1}{4}x + 2\right)^2$

11. Find the **area** of the rectangle below.

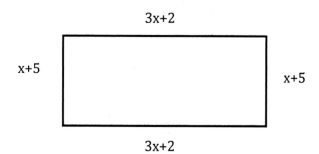

3x+2

x+5

x+5

3x+2

12. Find the **area** of the square below.

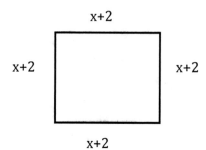

x+2

x+2

x+2

x+2

13. The **perimeter** of the square below is 60 cm. Find x.

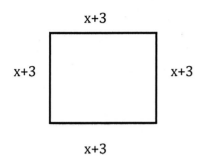

For more practice, access the online homework. See your syllabus for details.

Section 3.4 – Dividing Polynomials (Long Division)

❖ Dividing polynomials

❖ *Dividing polynomials*

In this section, we will divide polynomials. When dividing polynomials, we do have a few options. If possible, we should try to **factor** the numerator and denominator and cancel the common factors.

Also, we can use either synthetic division or long division. **Synthetic division** is a shortcut method and it can be used with factorable and non-factorable polynomials. However, it has its limitation because it can only be used when dividing by a linear factor $(ax + b)$.

The **long division** method is complete; it can be used with factorable and non-factorable polynomials, with linear and nonlinear factors. This is the one that will be discussed in this section.

In the other sections, I have attempted to provide the explanations and/or steps first then show a few examples. For this section, it is easier to provide the explanations as we work out the examples. Below are a few examples showing how to use long division when dividing polynomials.

Example 1

Divide.

$(x^2 + 3x - 4) \div (x - 1)$

This can be solved without long division. Using factoring, we have:

$$\frac{x^2+3x-4}{x-1} = \frac{(x-1)(x+4)}{x-1} = x + 4 \qquad \text{(Cancel } x - 1\text{)}$$

Using **long division**:

Make sure that all the polynomials are written in descending order.

- The divisor (denominator) goes to the left of the symbol and the dividend (numerator) goes to the right of the symbol.

$$x-1 \overline{\smash{)}\,x^2 + 3x - 4}$$

- Pay attention to the leading term x on the left and the leading term x^2 on the right. Divide x^2 by x. Write the result on top of the symbol.

$$x-1 \overline{\smash)x^2 + 3x - 4} \quad \overset{x}{}$$

- Multiply all the terms of the divisor $(x - 1)$ by x, change to opposite signs, and write the results underneath.

$$
\begin{array}{r}
x \phantom{{}+3x-4} \\
x-1 \overline{\smash)x^2 + 3x - 4} \\
\underline{-x^2 + x} \phantom{{}-4}
\end{array}
$$

- Add them up. The first term should cancel out. Bring down the next term.

$$
\begin{array}{r}
x \phantom{{}+3x-4} \\
x-1 \overline{\smash)x^2 + 3x - 4} \\
\underline{-x^2 + x} \phantom{{}} \downarrow \\
4x - 4
\end{array}
$$

- Pay attention to the leading term x on the left and the new leading term $4x$ on the right. Divide $4x$ by x. Write the result on top of the symbol next to x.

$$
\begin{array}{r}
x + 4 \phantom{{}-4} \\
x-1 \overline{\smash)x^2 + 3x - 4} \\
\underline{-x^2 + x} \phantom{{}} \\
4x - 4
\end{array}
$$

- Multiply all the terms of the divisor $(x - 1)$ by 4, change to opposite signs, and write the results underneath.

$$
\begin{array}{r}
x + 4 \phantom{{}-4} \\
x-1 \overline{\smash)x^2 + 3x - 4} \\
\underline{-x^2 + x} \phantom{{}} \\
4x - 4 \\
\underline{-4x + 4}
\end{array}
$$

- Add them up. The first term should cancel out. In this case, all the terms cancel out.

$$\begin{array}{r} x+4 \\ x-1\overline{\smash{)}x^2+3x-4} \\ \underline{-x^2+x} \\ 4x-4 \\ \underline{-4x+4} \\ 0 \end{array}$$

The answer is $x + 4$. There is no remainder.

Example 2

Divide.

$(2x^2 + 3x + 7) \div (3 + x)$

Using **long division**:

Make sure that all the polynomials are written in descending order.

- The divisor goes to the left of the symbol and the dividend goes to the right of the symbol.

$$x+3\overline{\smash{)}2x^2+3x+7}$$

- Pay attention to the leading term x on the left and the leading term $2x^2$ on the right. Divide $2x^2$ by x. Write the result on top of the symbol.

$$\begin{array}{r} 2x \\ x+3\overline{\smash{)}2x^2+3x+7} \end{array}$$

- Multiply all the terms of the divisor $(x + 3)$ by $2x$, change to opposite signs, and write the results underneath.

$$\begin{array}{r} 2x \\ x+3\overline{\smash{)}2x^2+3x+7} \\ \underline{-2x^2-6x} \end{array}$$

- Add them up. The first term should cancel out. Bring down the next term.

$$
\begin{array}{r}
2x \\
x+3\overline{)2x^2+3x+7} \\
\underline{-2x^2-6x} \qquad\downarrow \\
-3x+7
\end{array}
$$

- Pay attention to the leading term x on the left and the **new** leading term $-3x$ on the right. Divide $-3x$ by x. Write the result on top of the symbol next to $2x$.

$$
\begin{array}{r}
2x-3 \\
x+3\overline{)2x^2+3x+7} \\
\underline{-2x^2-6x} \\
-3x+7
\end{array}
$$

- Multiply all the terms of the divisor $(x+3)$ by -3, change to opposite signs, and write the results underneath.

$$
\begin{array}{r}
2x-3 \\
x+3\overline{)2x^2+3x+7} \\
\underline{-2x^2-6x} \\
-3x+7 \\
\underline{3x+9}
\end{array}
$$

- Add them up. The first term should cancel out.

$$
\begin{array}{r}
2x-3 \\
x+3\overline{)2x^2+3x+7} \\
\underline{-2x^2-6x} \\
-3x+7 \\
\underline{3x+9} \\
16
\end{array}
$$

One cannot go any further. The answer is $2x - 3 + \dfrac{16}{x+3}$.

Example 3

Divide.

$$(5x^3 + 4x - 7x^2) \div (x^2 + 2x + 1)$$

Using **long division**:

Make sure that all the polynomials are written in descending order.

- The divisor goes to the left of the symbol and the dividend goes to the right of the symbol.

$$x^2 + 2x + 1 \overline{\smash{)}5x^3 - 7x^2 + 4x}$$

- Pay attention to the leading term x^2 on the left and the leading term $5x^3$ on the right. Divide $5x^3$ by x^2. Write the result on top of the symbol.

$$\begin{array}{r} 5x \\ x^2 + 2x + 1 \overline{\smash{)}5x^3 - 7x^2 + 4x} \end{array}$$

- Multiply all the terms of the divisor $(x^2 + 2x + 1)$ by $5x$, change to opposite signs, and write the results underneath.

$$\begin{array}{r} 5x \\ x^2 + 2x + 1 \overline{\smash{)}5x^3 - 7x^2 + 4x} \\ -5x^3 - 10x^2 - 5x \end{array}$$

- Add them up. The first term should cancel out.

$$\begin{array}{r} 5x \\ x^2 + 2x + 1 \overline{\smash{)}5x^3 - 7x^2 + 4x} \\ -5x^3 - 10x^2 - 5x \\ \hline -17x^2 - x \end{array}$$

- Pay attention to the leading term x^2 on the left and the new leading term $-17x^2$ on the right. Divide $-17x^2$ by x^2. Write the result on top of the symbol next to $5x$.

$$\begin{array}{r} 5x - 17 \\ x^2 + 2x + 1 \overline{\smash{)}5x^3 - 7x^2 + 4x} \\ -5x^3 - 10x^2 - 5x \\ \hline -17x^2 - x \end{array}$$

- Multiply all the terms of the divisor $(x^2 + 2x + 1)$ by -17, change to opposite signs, and write the results underneath.

$$
\begin{array}{r}
5x - 17 \\
x^2 + 2x + 1{\overline{\smash{\big)}\,5x^3 - 7x^2 + 4x}} \\
\underline{-5x^3 - 10x^2 - 5x} \\
-17x^2 - x \\
\underline{+17x^2 + 34x + 17}
\end{array}
$$

- Add them up. The first term should cancel out.

$$
\begin{array}{r}
5x - 17 \\
x^2 + 2x + 1{\overline{\smash{\big)}\,5x^3 - 7x^2 + 4x}} \\
\underline{-5x^3 - 10x^2 - 5x} \\
-17x^2 - x \\
\underline{+17x^2 + 34x + 17} \\
33x + 17
\end{array}
$$

One cannot go any further. The answer is $5x - 17 + \dfrac{33x+17}{x^2+2x+1}$.

Example 4

Divide.

$(x^4 + 10x^2 - x + 9) \div (x^2 - 5x + 3)$

Using **long division**:

Make sure that all the polynomials are written in descending order. Use coefficient 0 for any missing term.

- The divisor goes to the left of the symbol and the dividend goes to the right of the symbol.

$$x^2 - 5x + 3{\overline{\smash{\big)}\,x^4 + 0x^3 + 10x^2 - x + 9}}$$

- Pay attention to the leading term x^2 on the left and the leading term x^4 on the right. Divide x^4 by x^2. Write the result on top of the symbol.

$$
\begin{array}{r}
x^2 \\
x^2 - 5x + 3{\overline{\smash{\big)}\,x^4 + 0x^3 + 10x^2 - x + 9}}
\end{array}
$$

- Multiply all the terms of the divisor $(x^2 - 5x + 3)$ by x^2, change to opposite signs, and write the results underneath.

$$x^2 - 5x + 3 \overline{\smash{)}\,x^4 + 0x^3 + 10x^2 - x + 9} \quad \overset{x^2}{}$$
$$\underline{-x^4 + 5x^3 - 3x^2}$$

- Add them up. The first term should cancel out. Bring down the next term.

$$x^2 - 5x + 3 \overline{\smash{)}\,x^4 + 0x^3 + 10x^2 - x + 9} \quad \overset{x^2}{}$$
$$\underline{-x^4 + 5x^3 - 3x^2} \quad \downarrow$$
$$5x^3 + 7x^2 - x$$

- Pay attention to the leading term x^2 on the left and the new leading term $5x^3$ on the right. Divide $5x^3$ by x^2. Write the result on top of the symbol next to x^2.

$$x^2 - 5x + 3 \overline{\smash{)}\,x^4 + 0x^3 + 10x^2 - x + 9} \quad \overset{x^2 + 5x}{}$$
$$\underline{-x^4 + 5x^3 - 3x^2}$$
$$5x^3 + 7x^2 - x$$

- Multiply all the terms of the divisor $(x^2 - 5x + 3)$ by $5x$, change to opposite signs, and write the results underneath.

$$x^2 - 5x + 3 \overline{\smash{)}\,x^4 + 0x^3 + 10x^2 - x + 9} \quad \overset{x^2 + 5x}{}$$
$$\underline{-x^4 + 5x^3 - 3x^2}$$
$$5x^3 + 7x^2 - x$$
$$\underline{-5x^3 + 25x^2 - 15x}$$

- Add them up. The first term should cancel out. Bring down the last term.

$$x^2 - 5x + 3 \overline{\smash{)}\,x^4 + 0x^3 + 10x^2 - x + 9} \quad \overset{x^2 + 5x}{}$$
$$\underline{-x^4 + 5x^3 - 3x^2}$$
$$5x^3 + 7x^2 - x$$
$$\underline{-5x^3 + 25x^2 - 15x} \quad \downarrow$$
$$32x^2 - 16x + 9$$

- Pay attention to the leading term x^2 on the left and the new leading term $32x^2$ on the right. Divide $32x^2$ by x^2. Write the result on top of the symbol next to $x^2 + 5x$.

$$
\begin{array}{r}
x^2 + 5x + 32 \\
x^2 - 5x + 3{\overline{\smash{\big)}\,x^4 + 0x^3 + 10x^2 - x + 9}} \\
\underline{-x^4 + 5x^3 - 3x^2} \\
5x^3 + 7x^2 - x \\
\underline{-5x^3 + 25x^2 - 15x} \\
32x^2 - 16x + 9
\end{array}
$$

- Multiply all the terms of the divisor $(x^2 - 5x + 3)$ by 32, change to opposite signs, and write the results underneath.

$$
\begin{array}{r}
x^2 + 5x + 32 \\
x^2 - 5x + 3{\overline{\smash{\big)}\,x^4 + 0x^3 + 10x^2 - x + 9}} \\
\underline{-x^4 + 5x^3 - 3x^2} \\
5x^3 + 7x^2 - x \\
\underline{-5x^3 + 25x^2 - 15x} \\
32x^2 - 16x + 9 \\
\underline{-32x^2 + 160x - 96}
\end{array}
$$

- Add them up. The first term should cancel out.

$$
\begin{array}{r}
x^2 + 5x + 32 \\
x^2 - 5x + 3{\overline{\smash{\big)}\,x^4 + 0x^3 + 10x^2 - x + 9}} \\
\underline{-x^4 + 5x^3 - 3x^2} \\
5x^3 + 7x^2 - x \\
\underline{-5x^3 + 25x^2 - 15x} \\
32x^2 - 16x + 9 \\
\underline{-32x^2 + 160x - 96} \\
144x - 87
\end{array}
$$

One cannot go any further. The answer is $x^2 + 5x + 32 + \dfrac{144x - 87}{x^2 - 5x + 3}$

Section 3.4 – Dividing polynomials (Long division)

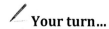 **Your turn...**

Divide.

1. $(x^2 - x - 42) \div (x + 6)$

2. $(x^2 - x - 2) \div (x + 1)$

3. $(x^2 + 16x + 63) \div (x + 7)$

4. $(2x^2 - 21x - 11) \div (x - 11)$

5. $(4x^2 + 14 + 6x) \div (x + 5)$

6. $(3x^3 - 5x^2 + 4) \div (x^2 + 7 + 4x)$

7. $(x^4 + 7x^2 - 2x + 11) \div (x^2 - 6x + 5)$

For more practice, access the online homework. See your syllabus for details.

Section 3.5 – Factoring Polynomials

- ❖ Greatest Common Factor (GCF)
- ❖ Binomial
- ❖ Grouping
- ❖ Trinomial, $a = 1$
- ❖ Trinomial, $a \neq 1$
- ❖ Perfect square trinomial
- ❖ Difference of two squares
- ❖ Combination of methods

Factoring can be explained as the reverse of the distributive property and FOIL. For example, use FOIL to multiply the binomials $(x + 2)(x + 3)$, the result is $x^2 + 5x + 6$. If given the trinomial $x^2 + 5x + 6$, the result after factoring, is $(x + 2)(x + 3)$. *See the chart at the end of this section for a summary of the factoring methods.*

In this section, we will concentrate on factoring the following:

❖ *Greatest Common Factor (GCF)*

Find the term that is common to all terms or GCF, factor it out, and express the result as a product by writing down the terms that are left.

Example 1

Factor.

$2x^2 + 10y^3 - 8z^5$ (The number 2 is common to all terms.)

The GCF is 2, so factor out the 2.

$2(x^2 + 5y^3 - 4z^5)$

To check if the final answer is correct, distribute the 2, and the result is the original expression.

Example 2

Factor.

$24m^6n^5 + 15m^2n^3$ (Factor 3 and the lowest powers of m and n.)

The GCF is $3m^2n^3$, so factor out the $3m^2n^3$.

$3m^2n^3(8m^4n^2 + 5)$

To check if the final answer is correct, distribute the $3m^2n^3$, and the result is the original expression.

Example 3

Factor.

$7ab^2c^5 - 28a^2bc^4 + 49abc^3$ (Factor 7 and the lowest powers of a, b, and c.)

The GCF is $7abc^3$, so factor out the $7abc^3$.

$7abc^3(bc^2 - 4ac + 7)$

To check if the final answer is correct, distribute the $7abc^3$, and the result is the original expression.

> ❖ *Binomial*

Find the binomial that is common to all the terms, factor it out, and express the result as a product by writing down the terms that are left.

Example 4

Factor.

$x^2(x + 5) + 3(x + 5)$ (The binomial $x + 5$ is common, factor it out.)

$(x + 5)(x^2 + 3)$

Example 5

Factor.

$y^3(7y - 4) + y(7y - 4) - 2(7y - 4)$ (The binomial $7y - 4$ is common, factor it out.)

$(7y - 4)(y^3 + y - 2)$

Example 6

Factor.

$\frac{1}{7}k\left(k + \frac{1}{4}\right) - 5\left(k + \frac{1}{4}\right)$ (The binomial $k + \frac{1}{4}$ is common, factor it out.)

$\left(k + \frac{1}{4}\right)\left(\frac{1}{7}k - 5\right)$

❖ *Grouping*

This method involves factoring out both a GCF and a binomial. Group the polynomial by factoring out a GCF from the first 2 terms and factoring out a GCF from the last 2 terms. A binomial is common on the next step. Factor out the common binomial and write down the terms that are left. The final answer is a product of 2 binomials.

Example 7

Factor.

$x^3 - 3x^2 + 4x - 12$ (Look at the first 2 terms, factor out a GCF; look at the last 2 terms, and do the same.)

$x^3 - 3x^2 \quad + 4x - 12$

$x^2(x - 3) \quad + 4(x - 3)$ (The binomial $x - 3$ is common, factor it out.)

$(x - 3)(x^2 + 4)$

To check if the final answer is correct, use FOIL, and the result is the original expression.

Example 8

Factor.

$2y^2 - 14y - 5y + 35$ (Look at the first 2 terms, factor out a GCF; look at the last 2 terms, and do the same.)

$2y^2 - 14y \quad - 5y + 35$

$2y(y - 7) \quad - 5(y - 7)$ (The binomial $y - 7$ is common, factor it out.)

$(y - 7)(2y - 5)$

To check if the final answer is correct, use FOIL, and the result is the original expression.

Example 9

Factor.

$14k^2 + 35k - 12k - 30$ (Look at the first 2 terms, factor out a GCF; look at the last 2 terms, and do the same.)

$14k^2 + 35k \quad - 12k - 30$

$7k(2k + 5) \quad - 6(2k + 5)$ (The binomial $2k + 5$ is common, factor it out.)

$(2k + 5)(7k - 6)$ To check if the final answer is correct, use FOIL, and the result is the original expression.

❖ *Trinomial*, $a = 1$ $(ax^2 + bx + c)$ or $(x^2 + bx + c)$

If the coefficient of x^2 is 1, we will only use the last term c. Find 2 numbers that when multiplied result in c and when added, result in the coefficient of the middle term, b. Once the numbers found satisfy these criteria, use them to factor the trinomial as the product of 2 binomials.

Example 10

Factor.

$x^2 + 7x + 10$

$a = 1, \ b = 7, \ c = 10$

$x^2 + 7x + \mathbf{10}$ (2 and 5 both positive; $2 \cdot 5 = 10$; $2 + 5 = 7$)

$(x + 2)(x + 5)$ or $(x + 5)(x + 2)$

To check if the final answer is correct, use FOIL, and the result is the original expression.

Example 11

Factor.

$x^2 - 2x - 63$

$a = 1, \ b = -2, c = -63$

$x^2 - 2x - \mathbf{63}$ (-9 and 7; $-9 \cdot 7 = -63$; $-9 + 7 = -2$)

$(x + 7)(x - 9)$ or $(x - 9)(x + 7)$

To check if the final answer is correct, use FOIL, and the result is the original expression.

Example 12

Factor.

$x^2 - 8x + 12$

$a = 1, \ b = -8, c = 12$

$x^2 - 8x + \mathbf{12}$ (-6 and -2; $-6 \cdot -2 = 12$; $-6 + (-2) = -8$)

$(x - 6)(x - 2)$ or $(x - 2)(x - 6)$

To check if the final answer is correct, use FOIL, and the result is the original expression.

Example 13

Factor.

$x^2 + 6x + 9$

$a = 1, \ b = 6, \ c = 9$

$x^2 + 6x + \mathbf{9}$ (3 and 3; $3 \cdot 3 = 9$; $3 + 3 = 6$)

$(x + 3)(x + 3)$ or $(x + 3)^2$

To check if the final answer is correct, use FOIL, and the result is the original expression.

❖ **Trinomial, $a \neq 1 \ (ax^2 + bx + c)$**

There are many ways to factor the trinomial when $a \neq 1$. In this book, we discuss only one method. It is by grouping. Multiply the coefficient of the first term, a, with the last term, c. Find 2 numbers when multiplied, the result is ac, and when added, the result is the coefficient of the middle term, b. Keep the first term and last term of the trinomial as is. Rewrite the middle term using the 2 numbers found. The result is a polynomial of four terms. Factor the new polynomial by grouping, discussed earlier in this section. This method is FOIL backward.

Example 14

Factor.

$2x^2 + 7x + 6$

$a = 2, \ b = 7, \ c = 6$

$2x^2 + \mathbf{7}x + 6$ ($2 \cdot 6 = 12$; the two numbers are 4 and 3; $4 \cdot 3 = 12$; $4 + 3 = 7$)

$2x^2 + \mathbf{4}x + \mathbf{3}x + 6$ (Rewrite the middle term using 4 and 3– order does not matter.)

$2x^2 + 4x \quad + 3x + 6$ (Factor by grouping.)

$2x(x + 2) \quad + 3(x + 2)$ (The binomial $x + 2$ is common, factor it out.)

$(x + 2)(2x + 3)$ or $(2x + 3)(x + 2)$

To check if the final answer is correct, use FOIL, and the result is the original expression.

Example 15

Factor.

$6x^2 - 13x + 5$

$a = 6, \ b = -13, \ c = 5$

$6x^2 - 13x + 5$ ($6 \cdot 5 = 30$; the numbers are -3 and -10; $(-3) \cdot (-10) = 30$; $-3 + (-10) = -13$)

$6x^2 - 3x - 10x + 5$ (Rewrite the middle term using -3 and -10 – order does not matter.)

$6x^2 - 3x \quad - 10x + 5$ (Factor by grouping.)

$3x(2x - 1) \ - 5(2x - 1)$ (The binomial $2x - 1$ is common, factor it out.)

$(2x - 1)(3x - 5)$ or $(3x - 5)(2x - 1)$

To check if the final answer is correct, use FOIL, and the result is the original expression.

Example 16

Factor.

$4x^2 - 4x - 15$

$a = 4, \ b = -4, \ c = -15$

$4x^2 - 4x - 15$ ($4 \cdot -15 = -60$; the numbers are -10 and 6; $(-10) \cdot (6) = -60$; $-10 + 6 = -4$)

$4x^2 - 10x + 6x - 15$ (Rewrite the middle term using -10 and 6 – order does not matter.)

$4x^2 - 10x \quad + 6x - 15$ (Factor by grouping.)

$2x(2x - 5) \ + 3(2x - 5)$ (The binomial $2x - 5$ is common, factor it out.)

$(2x - 5)(2x + 3)$ or $(2x + 3)(2x - 5)$

To check if the final answer is correct, use FOIL, and the result is the original expression.

Example 17

Factor.

$3k^2 + k + 7$

$a = 3,\ b = 1,\ c = 7$

This is a prime trinomial. It cannot be factored. Two numbers are needed which product is 21 (ac) and which sum is 1 (b). No such numbers exist.

❖ *Perfect square trinomial*

To factor a perfect square trinomial, look at the first term and the last term. They both must be perfect squares. The middle term must be the product of 2, the square root of the first term, and the square root of the last term. If this is the case, then use the square root of the first and last term when writing the factors of the two binomials and use the sign of the middle term.

$$a^2 \pm 2ab + b^2 = (a \pm b)^2$$

Example 18

Factor.

$4x^2 + 12x + 9$

The first and the last term are perfect squares: $(2x)^2$ and $(3)^2$ and the middle term is the product of $2(2x)(3) = 12x$. Then use the square root of the first and last term to factor the trinomial. Use the sign of the middle term.

The result is: $(2x + 3)(2x + 3)$ or $(2x + 3)^2$

To check if the final answer is correct, use FOIL, and the result is the original expression.

Example 19

Factor.

$9x^2 - 6x + 1$

The first and the last term are perfect squares: $(3x)^2$ and $(1)^2$ and the middle term is the product of $2(3x)(1) = 6x$. Then use the square root of the first and last term to factor the trinomial. Use the sign of the middle term.

The result is: $(3x - 1)(3x - 1)$ or $(3x - 1)^2$

To check if the final answer is correct, use FOIL, and the result is the original expression.

Example 20

Factor.

$16x^2 + 40x + 25$

The first and the last term are perfect squares: $(4x)^2$ and $(5)^2$ and the middle term is the product of $2(4x)(5) = 40x$. Then use the square root of the first and last term to factor the trinomial. Use the sign of the middle term.

The result is: $(4x + 5)(4x + 5)$ or $(4x + 5)^2$

To check if the final answer is correct, use FOIL, and the result is the original expression.

❖ *Difference of two squares*

The difference of two squares describes a binomial that has a minus sign between the terms and each term is a perfect square. If all these criteria are met, write the square roots of the perfect squares as a product of 2 binomials with a minus sign for one of the binomials and a plus sign for the other binomial.

An example of this is: $x^2 - 144 = (x - 12)(x + 12)$.

We can also use the formula:

$a^2 - b^2 = (a + b)(a - b)$

Example 21

Factor.

$81x^2 - 25y^2$

$(9x + 5y)(9x - 5y)$ or $(9x - 5y)(9x + 5y)$

To check if the final answer is correct, use FOIL, and the result is the original expression.

Example 22

Factor.

$\frac{4}{16}x^2 - \frac{9}{25}$

$\left(\frac{2}{4}x + \frac{3}{5}\right)\left(\frac{2}{4}x - \frac{3}{5}\right)$ or $\left(\frac{2}{4}x - \frac{3}{5}\right)\left(\frac{2}{4}x + \frac{3}{5}\right)$

To check if the final answer is correct, use FOIL, and the result is the original expression.

Example 23

Factor.

$9k^2 + 49$

This is a prime binomial. This is not the **difference** of 2 squares. It cannot be factored using any other method either.

Example 24

Factor.

$64 - y^2$

$(8 - y)(8 + y)$ or $(8 + y)(8 - y)$

To check if the final answer is correct, use FOIL, and the result is the original expression.

Example 25

Factor.

$u^2 - v^2$

$(u - v)(u + v)$ or $(u + v)(u - v)$

To check if the final answer is correct, use FOIL, and the result is the original expression.

Example 26

Factor.

$x^4 - 81$

$(x^2 - 9)(x^2 + 9)$

$(x - 3)(x + 3)(x^2 + 9)$

To check if the final answer is correct, use FOIL, and the result is the original expression.

❖ *Combination of methods*

Example 27

Factor.

$27y^2 + 54y + 27$

The numbers are large. Try to factor out a GCF first if possible before factoring.

$27\,(y^2 + 2y + 1)$ (Factor out a GCF first – 27 is common to all terms.)

$a = 1,\ b = 2,\ c = 1$

$27\,(y^2 + 2y + \mathbf{1})$ (Inside the parenthesis $1 \cdot 1 = 1$; The numbers are 1 and 1; $1 \cdot 1 = 1$; $1 + 1 = 2$)

$27\,(y + 1)(y + 1)$

To check if the final answer is correct, use FOIL, then distribute the 27, and the result is the original expression.

Example 28

Factor.

$48x^3y - 27xy^3$

This expression does not meet the criteria of the difference of 2 squares. Try to factor out a GCF first.

$48x^3y - 27xy^3$ (Factor 3 and the lowest powers of x and y.)

$3xy\,(16x^2 - 9y^2)$ (Inside the parenthesis is the difference of 2 squares.)

$3xy\,(4x + 3y)(4x - 3y)$

To check if the final answer is correct, use FOIL then distribute $3xy$, and the result is the original expression.

Rewrite Polynomial in Descending Order

$$-2 - 6x + x^2 = x^2 - 6x - 2$$

GCF

Factor out -1 if leading coefficient is negative

$$-x^2 + 4x - 3 = -1(x^2 - 4x + 3)$$

Don't forget!

How many terms in the polynomial?

2 terms

Try Difference of Squares:

$$a^2 - b^2 = (a + b)(a - b)$$

4 terms

Try Grouping

$$ax + ay \quad + bx + by$$

$$a(x + y) \quad + b(x + y)$$

$$(x + y)(a + b)$$

3 terms

Which type of trinomial is it?

$$ax^2 + bx + c, \quad a = 1$$

$(x \quad)(x \quad) =$ Look for factors of c that add to b. Check the signs!

$$ax^2 + bx + c, \qquad a \neq 1$$

AC or Grouping method: Look for factors of a times c that add to b. Keep first and last term. Rewrite the middle term, b, using factors of ac. The result is a polynomial of four terms, then factor by grouping.

Modified from Hector Alfaro's chart

Perfect Square Trinomial

$$a^2 \pm 2ab + b^2 = (a \pm b)^2$$

Section 3.5 – Factoring polynomials

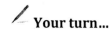 **Your turn...**

Factor.

1. $6x^2 + 30y^4 - 48z^5$

2. $12m^6n^7 + 15m^2n^3$

3. $14ab^3c^5 - 28a^2bc^4 + 63abc^3$

4. $x^2(x + 4) + 5(x + 4)$

5. $y^3(3y - 4) + y(3y - 4) - 5(3y - 4)$

6. $\frac{1}{7}k\left(k + \frac{1}{2}\right) - 9\left(k + \frac{1}{2}\right)$

7. $x^2 - 3x + 5x - 15$

8. $2y^2 - 14y - 3y + 21$

9. $7k^2 + 21k - 6k - 18$

10. $x^2 + 6x + 8$

11. $x^2 - x - 56$

12. $x^2 - 9x + 14$

13. $x^2 + 12x + 36$

14. $2x^2 + 7x + 3$

15. $6x^2 - 17x + 7$

16. $4x^2 - x - 3$

17. $15y^2 + 30y + 15$

18. $4k^2 + k + 9$

19. $49x^2 - 16y^2$

20. $\frac{4}{25}x^2 - \frac{16}{121}$

21. $32x^3y - 18xy^3$

22. $4k^2 + 81$

23. $36 - y^2$

24. $c^2 - d^2$

25. $9x^2 - 12x + 4$

26. $4x^2 + 20x + 25$

27. $36x^2 + 84x + 49$

28. $x^4 - 16$

For more practice, access the online homework. See your syllabus for details.

Section 3.6 – Factoring Polynomials: Sum and Difference of Cubes

- ❖ Difference of cubes
- ❖ Sum of cubes

When factoring the sum or difference of two cubes, use the following formulas:

$$a^3 - b^3 = (a - b)(a^2 + ab + b^2) \rightarrow \text{Difference of cubes}$$

$$a^3 + b^3 = (a + b)(a^2 - ab + b^2) \rightarrow \text{Sum of cubes}$$

Note that the binomials have the same signs. However, the middle terms of the trinomials have opposite signs when compared to the binomials. The trinomials of the formulas are not factorable.

❖ *Difference of cubes*

Example 1

Factor

$x^3 - 64$

$(x)^3 - (4)^3$ ($x = a$ and $4 = b$ when compared to the formula)

Using the formula:

$$a^3 - b^3 = (a - b)(a^2 + ab + b^2)$$

$$x^3 - 4^3 = (x - 4)(x^2 + x \cdot 4 + 4^2) = (x - 4)(x^2 + 4x + 16)$$

$$\therefore x^3 - 64 = (x - 4)(x^2 + 4x + 16)$$

To check if the answer is correct, multiply, and the result is the original expression.

Example 2

Factor

$8x^3 - 27$

$(2x)^3 - (3)^3$ ($2x = a$ and $3 = b$ when compared to the formula)

Using the formula:

$a^3 - b^3 = (a - b)(a^2 + ab + b^2)$

$(2x)^3 - 3^3 = (2x - 3)[(2x)^2 + 2x \cdot 3 + 3^2] = (2x - 3)(4x^2 + 6x + 9)$

$\therefore 8x^3 - 27 = (2x - 3)(4x^2 + 6x + 9)$

To check if the answer is correct, multiply, and the result is the original expression.

Example 3

Factor

$4x^3 - 32$

$4(x^3 - 8)$ (Factor a GCF first since the numbers 4 and 32 are not perfect cubes.)

$4[(x)^3 - (2)^3]$ ($x = a$ and $2 = b$ when compared to the formula)

Using the formula:

$a^3 - b^3 = (a - b)(a^2 + ab + b^2)$

$(x)^3 - 2^3 = (x - 2)[x^2 + x \cdot 2 + 2^2] = (x - 2)(x^2 + 2x + 4)$

Do not forget the 4.

$\therefore 4x^3 - 32 = 4(x - 2)(x^2 + 2x + 4)$

To check if the answer is correct, multiply, and the result is the original expression.

❖ *Sum of cubes*

Example 4

Factor

$x^3 + 1$

$(x)^3 + (1)^3$ $\qquad$ ($x = a$ and $1 = b$ when compared to the formula)

Using the formula:

$a^3 + b^3 = (a + b)(a^2 - ab + b^2)$

$x^3 + 1^3 = (x + 1)(x^2 - x \cdot 1 + 1^2) = (x + 1)(x^2 - x + 1)$

$\therefore x^3 + 1 = (x + 1)(x^2 - x + 1)$

To check if the answer is correct, multiply, and the result is the original expression.

Example 5

Factor

$y^3 + 125$

$(y)^3 + (5)^3$ $\qquad$ ($y = a$ and $5 = b$ when compared to the formula)

Using the formula:

$a^3 + b^3 = (a + b)(a^2 - ab + b^2)$

$y^3 + 5^3 = (y + 5)(y^2 - y \cdot 5 + 5^2) = (y + 5)(y^2 - 5y + 25)$

$\therefore y^3 + 125 = (y + 5)(y^2 - 5y + 25)$

To check if the answer is correct, multiply, and the result is the original expression.

Section 3.6 – Factoring polynomials: sum and difference of cubes

✎ **Your turn...**

Factor.

1. $y^3 - 216$

2. $27y^3 - 8$

3. $2x^3 - 54$

4. $x^3 + 64$

5. $x^3 + 1$

6. $64y^3 + 125$

For more practice, access the online homework. See your syllabus for details.

Chapter 3: Review of Terms, Concepts, and Formulas

- A **polynomial** is an expression comprising of the sum of two or more terms, made of variables, constants, and nonnegative integer exponents.

- A **term** is the product of a coefficient and a variable or variables.

- The **coefficient** is the number in front of the variable.

- The **degree** of a term refers to the exponent or power of the term. The degree of the polynomial is the highest degree when comparing the degree of all the terms.

- An expression with one term is classified as a **monomial**.

- An expression with two terms is classified as a **binomial**.

- An expression with three terms is classified as a **trinomial**.

- The polynomial is in **descending order** when the degree of the terms starts from the highest to the lowest.

- The **leading** term is the term with the highest degree. The leading coefficient is the number in front of the variable of the leading term or the number in front of the variable with the highest degree.

- In order to **add or subtract polynomials**, the polynomial must contain like terms. In other words, the terms must have the same variables and degrees. If they are like terms, add the like terms by adding the coefficients. The variable never changes in addition or subtraction, but the coefficients change.

- To **multiply polynomials**, multiply the coefficients together and the variables together. If the terms have the same variable, keep the base or variable, and add the exponents. In the process of multiplying polynomials, we also use the distributive property and the FOIL method when applicable.

  We can also use the formulas below:
  - Sum and difference: $(a + b)(a - b) = a^2 - b^2$
  - Binomial squared: $(a + b)^2 = a^2 + 2ab + b^2$
  - Binomial squared: $(a - b)^2 = a^2 - 2ab + b^2$

- Factoring a **GCF** $\rightarrow ax + 2ay - 3az = a(x + 2y - 3z) \rightarrow$ the GCF is a.

- Factoring the **difference of 2 squares** $\rightarrow a^2 - b^2 = (a + b)(a - b)$

- Factoring by **grouping**:

$$ax + ay \ + \ bx + by$$

$$a(x + y) \ + \ b(x + y)$$

$$(x + y)(a + b)$$

- Factoring trinomial: $ax^2 + bx + c, \ a = 1$

  $(x \quad)(x \quad) =$ Look for factors of c that add to b. Check the signs.

- Factoring trinomial: $ax^2 + bx + c, \ a \neq 1$

  AC or grouping method: Look for factors of a times c that add to b. Keep first and last term. Rewrite the middle term, bx, using factors of ac. The result is a polynomial of four terms, then factor by grouping.

- Factoring a perfect square trinomial: $a^2 \pm 2ab + b^2 = (a \pm b)^2$

- Factoring the difference of cubes: $a^3 - b^3 = (a - b)(a^2 + ab + b^2)$

- Factoring the sum of cubes: $a^3 + b^3 = (a + b)(a^2 - ab + b^2)$

CHAPTER 4

RATIONAL EXPRESSIONS, EQUATIONS, AND FUNCTIONS

Section 4.1 - Simplifying, Finding Domains, and Evaluating

- ❖ Simplifying rational expressions
- ❖ Finding the domain of rational functions
- ❖ Evaluating rational functions

❖ ***Simplifying rational expressions (also called algebraic fractions)***

Examples of Rational Expressions

$$\frac{2x - 5}{x + 7}$$

$$\frac{6}{x^2 - x - 42}$$

$$\frac{x^3 + x}{x - 1}$$

To simplify rational expressions (expressions in ratio form with a variable in the denominator), factor all the expressions in the numerator and denominator if applicable. Then simplify any common factors from the numerator and denominator. Remember that only factors (multiplication) can be cancelled out (or reduced to 1).

Example 1

Simplify the rational expression.

$$\frac{3x + 21}{5x + 35}$$

$$\frac{3(x + 7)}{5(x + 7)} = \frac{3\cancel{(x + 7)}}{5\cancel{(x + 7)}} = \frac{3}{5}$$

Example 2

Simplify the rational expression.

$$\frac{x^2 - 2x - 3}{x^2 + 6x + 5}$$

$$\frac{x^2 - 2x - 3}{x^2 + 6x + 5} = \frac{(x - 3)(x + 1)}{(x + 5)(x + 1)} = \frac{(x - 3)\cancel{(x + 1)}}{(x + 5)\cancel{(x + 1)}} = \frac{x - 3}{x + 5}$$

Example 3

Simplify the rational expression.

$$\frac{2x^2 - 7x - 4}{x^2 + 3x - 28}$$

$$\frac{(2x + 1)(x - 4)}{(x + 7)(x - 4)} = \frac{(2x + 1)\cancel{(x - 4)}}{(x + 7)\cancel{(x - 4)}} = \frac{2x + 1}{x + 7}$$

Example 4

Simplify the rational expression.

$$\frac{x - 7}{49 - x^2}$$

$$\frac{x - 7}{(7 - x)(7 + x)} = \frac{-1(-x + 7)}{(7 - x)(7 + x)} = \frac{-1\cancel{(-x + 7)}}{\cancel{(7 - x)}(7 + x)} = \frac{-1}{7 + x}$$

Example 5

Simplify the rational expression.

$$\frac{16 - 2x}{7x - 56}$$

$$\frac{2(8 - x)}{7(x - 8)} = \frac{-2(-8 + x)}{7(x - 8)} = \frac{-2\cancel{(-8 + x)}}{7\cancel{(x - 8)}} = -\frac{2}{7}$$

Example 6

Simplify the rational expression.

$$\frac{5x^3 - 40}{3x^2 + 6x + 12}$$

$$\frac{5(x^3 - 8)}{3(x^2 + 2x + 4)} = \frac{5(x - 2)(x^2 + 2x + 4)}{3(x^2 + 2x + 4)} = \frac{5(x - 2)\cancel{(x^2 + 2x + 4)}}{3\cancel{(x^2 + 2x + 4)}} = \frac{5(x - 2)}{3} \text{ or } \frac{5x - 10}{3}$$

❖ *Finding the domain of rational functions*

In section 1.5, we discussed functions and their domains. In this section, we will discuss the domain of functions again and will concentrate only on rational functions.

Rational functions are in the form of a fraction or ratio and have a variable in the denominator. As with any fraction, there cannot be a zero in the denominator. Therefore, there cannot be a zero in the denominator of a rational function. To find the domain for this type of function, set the denominator equal to zero and solve it. The value(s) found, if any, will not be in the domain. In other words, the domain will be all real numbers except the value(s) found when solving the denominator of the function.

Example 7

Find the domain of the function.

$$A(x) = \frac{x+1}{x-2}$$

$x - 2 = 0 \rightarrow x = 2 \rightarrow x \neq 2$ (Set the denominator equal to 0 and solve it.)

The number 2 makes the denominator of the function zero. The domain is all real numbers except 2 or $\{x \mid x \neq 2\}$ or $(-\infty, 2) \cup (2, \infty)$

Example 8

Find the domain of the function.

$$B(x) = \frac{2x-5}{(x+7)(x-5)}$$

$x + 7 = 0 \rightarrow x = -7 \rightarrow x \neq -7$ (Set each factor in the denominator equal to 0 and solve it.)

$x - 5 = 0 \rightarrow x = 5 \rightarrow x \neq 5$ (Set each factor in the denominator equal to 0 and solve it.)

The numbers -7 and 5 make the denominator of the function zero. The domain is all real numbers except -7 and 5 or $\{x \mid x \neq -7, x \neq 5\}$ or $(-\infty, -7) \cup (-7, 5) \cup (5, \infty)$.

Example 9

Find the domain of the function.

$$C(x) = \frac{6}{x^2 - x - 72}$$

$x^2 - x - 72 = 0$ (Set the denominator equal to 0.)

$(x - 9)(x + 8) = 0$ (Factor.)

$x - 9 = 0 \rightarrow x = 9 \rightarrow x \neq 9$ (Solve each factor.)

$x + 8 = 0 \rightarrow x = -8 \rightarrow x \neq -8$ (Solve each factor.)

The numbers -8 and 9 make the denominator of the function zero. The domain is all real numbers except -8 and 9 or $\{x | x \neq -8, x \neq 9\}$ or $(-\infty, -8) \cup (-8, 9) \cup (9, \infty)$.

Example 10

Find the domain of the function.

$$D(x) = \frac{x}{x^2 + 1}$$

$x^2 + 1 = 0$ (Cannot be factored)

$x^2 = -1$ (Solve it for x.)

$x = \pm\sqrt{-1}$ (Not a real number)

The domain consists of **all real numbers** or $(-\infty, \infty)$. We cannot find any x-values that make the denominator zero. All x-values are in the domain or work for this function.

❖ *Evaluating rational functions*

Evaluating functions mean finding the y −value given an x −value and vice versa. Simply replace the given value in the function.

Example 11

Evaluate the rational function.

$$f(x) = \frac{x+1}{x+2}$$

Find $f(0), f(-1), f(2), f(-4),$ and x when $f(x) = 3$.

a) $f(0) = \frac{0+1}{0+2} = \frac{1}{2}$ (Replace x with 0.)

b) $f(-1) = \frac{-1+1}{-1+2} = \frac{0}{1} = 0$ (Replace x with -1.)

c) $f(2) = \frac{2+1}{2+2} = \frac{3}{4}$ (Replace x with 2.)

d) $f(-4) = \frac{-4+1}{-4+2} = \frac{-3}{-2} = \frac{3}{2}$ (Replace x with -4.)

e) $3 = \frac{x+1}{x+2}$ (Replace $f(x)$ with 3.)

    $\frac{3}{1} = \frac{x+1}{x+2}$ (Rewrite 3 as a fraction)

    $3(x+2) = 1(x+1)$ (Use cross-multiplication.)

    $3x + 6 = x + 1$ (Distribute the 3 and the 1.)

    $2x = -5$ (Subtract x and 6 from both sides.)

    $x = -\frac{5}{2}$ (Divide by 2.)

Section 4.1 - Simplifying, finding domains, and evaluating

Your turn...

Simplify the rational expressions.

1. $\dfrac{2x+14}{4x+28}$

2. $\dfrac{x^2+2x-8}{x^2+9x+20}$

3. $\dfrac{2x^2-5x-12}{x^2+2x-24}$

4. $\dfrac{x-9}{81-x^2}$

5. $\dfrac{24-3x}{5x-40}$

6. $\dfrac{2y^2-10y+50}{3y^3+375}$

Find the domain of the functions.

7. $A(x) = \dfrac{x+1}{x-5}$

8. $B(x) = \dfrac{2x-5}{(x+3)(x-2)}$

9. $C(x) = \dfrac{6}{x^2-x-20}$

10. $D(x) = \dfrac{-3x+7}{x^2+36}$

Evaluate the rational function.

11. $f(x) = \dfrac{x}{x+2}$

Find $f(0), f(4), f(-5),$ and x when $f(x) = 6.$

For more practice, access the online homework. See your syllabus for details.

> **Section 4.2 – Multiplying & Dividing Rational Expressions**
>
> - Multiplying rational expressions
> - Dividing rational expressions

❖ *Multiplying rational expressions*

To multiply rational expressions, simplify the rational expressions first by factoring all the expressions. Simplify any common factors from the numerator and denominator. Then multiply the results.

Example 1

Multiply the rational expressions.

$$\frac{x^2 + 7x}{x^2} \cdot \frac{4}{x^2 - 49}$$

Factor each expression if possible and cancel common factors.

$$\frac{x(x + 7)}{x^2} \cdot \frac{4}{(x + 7)(x - 7)} = \frac{x\,\cancel{(x+7)}}{\cancel{x}\ x} \cdot \frac{4}{\cancel{(x+7)}(x - 7)}$$

$$\frac{1}{x} \cdot \frac{4}{(x - 7)} = \frac{4}{x(x - 7)} \text{ or } \frac{4}{x^2 - 7x}$$

Example 2

Multiply the rational expressions.

$$\frac{k^3 - 3k^2 - 4k}{k^2 - 1} \cdot \frac{9k - 9}{k^2 - 9k + 20}$$

Factor each expression if possible and cancel common factors.

$$\frac{k(k^2 - 3k - 4)}{(k + 1)(k - 1)} \cdot \frac{9(k - 1)}{(k - 5)(k - 4)}$$

$$\frac{k(k - 4)(k + 1)}{(k + 1)(k - 1)} \cdot \frac{9(k - 1)}{(k - 5)(k - 4)} = \frac{k\,\cancel{(k-4)}\cancel{(k+1)}}{\cancel{(k+1)}\cancel{(k-1)}} \cdot \frac{9\,\cancel{(k-1)}}{(k - 5)\cancel{(k-4)}}$$

$$\frac{9k}{(k - 5)}$$

Example 3

Multiply the rational expressions.

$$\frac{x^2 - 64}{3x + 27} \cdot \frac{10x}{5x + 40} \cdot \frac{x + 9}{x^2 - 8x}$$

Factor each expression if possible and cancel common factors.

$$\frac{(x - 8)(x + 8)}{3(x + 9)} \cdot \frac{10x}{5(x + 8)} \cdot \frac{x + 9}{x(x - 8)} = \frac{\cancel{(x - 8)}\cancel{(x + 8)}}{3\cancel{(x + 9)}} \cdot \frac{10\cancel{x}}{5\cancel{(x + 8)}} \cdot \frac{\cancel{x + 9}}{\cancel{x}\cancel{(x - 8)}}$$

$$\frac{10}{15} = \frac{2}{3}$$

Example 4

Multiply the rational expressions.

$$\frac{x^3 - 64}{x^2 - 16} \cdot \frac{x + 4}{5}$$

Factor each expression if possible and cancel common factors.

$$\frac{(x - 4)(x^2 + 4x + 16)}{(x - 4)(x + 4)} \cdot \frac{x + 4}{5}$$

$$\frac{\cancel{(x - 4)}(x^2 + 4x + 16)}{\cancel{(x - 4)}\cancel{(x + 4)}} \cdot \frac{\cancel{x + 4}}{5}$$

$$\frac{x^2 + 4x + 16}{5}$$

❖ *Dividing rational expressions*

To divide rational expressions, keep the first rational expression as is and multiply it by the reciprocal of the second rational expression. Factor all the expressions. Simplify any common factors from the numerator and denominator. Multiply the results.

Example 5

Divide the rational expressions.

$$\frac{x^3 - 9x}{x^2 - 5x} \div \frac{x^2 - 9}{x^2 + x - 30}$$

Multiply the first rational expression by the reciprocal of the second rational expression.

$$\frac{x^3 - 9x}{x^2 - 5x} \cdot \frac{x^2 + x - 30}{x^2 - 9}$$

Factor each expression if possible and cancel common factors.

$$\frac{x(x^2 - 9)}{x(x - 5)} \cdot \frac{(x + 6)(x - 5)}{(x^2 - 9)}$$

$$\frac{\cancel{x}\ \cancel{(x^2 - 9)}}{\cancel{x}\ \cancel{(x - 5)}} \cdot \frac{(x + 6)\cancel{(x - 5)}}{\cancel{(x^2 - 9)}}$$

$$x + 6$$

Example 6

Divide the rational expressions.

$$\frac{3x^2 - 13x + 4}{x^2 - 16} \div \frac{27x - 9}{2x + 8}$$

Multiply the first rational expression by the reciprocal of the second rational expression.

$$\frac{3x^2 - 13x + 4}{x^2 - 16} \cdot \frac{2x + 8}{27x - 9}$$

Factor each expression if possible and cancel common factors.

$$\frac{(x - 4)(3x - 1)}{(x - 4)(x + 4)} \cdot \frac{2(x + 4)}{9(3x - 1)} = \frac{\cancel{(x - 4)}\cancel{(3x - 1)}}{\cancel{(x - 4)}2\cancel{(x + 4)}} \cdot \frac{2\cancel{(x + 4)}}{9\cancel{(3x - 1)}}$$

$$\frac{2}{9}$$

Example 7

Divide the rational expressions.

$$\frac{18 - 2x}{7x^2 - 77x} \div \frac{2x^2 - 162}{7x}$$

Multiply the first rational expression by the reciprocal of the second rational expression.

$$\frac{18 - 2x}{7x^2 - 77x} \cdot \frac{7x}{2x^2 - 162}$$

Factor each expression if possible and cancel common factors.

$$\frac{2(9 - x)}{7x(x - 11)} \cdot \frac{7x}{2(x^2 - 81)}$$

$$\frac{2(9 - x)}{7x(x - 11)} \cdot \frac{7x}{2(x + 9)(x - 9)}$$

$$\frac{-2(-9 + x)}{7x(x - 11)} \cdot \frac{7x}{2(x + 9)(x - 9)} = \frac{-\cancel{2}(\cancel{-9 + x})}{\cancel{7x}(x - 11)} \cdot \frac{\cancel{7x}}{\cancel{2}(x + 9)\cancel{(x - 9)}}$$

$$\frac{-1}{(x - 11)(x + 9)}$$

Example 8

Divide the rational expressions.

$$\frac{x^3 + 1}{x^2 - 5x} \div \frac{x^2 - 1}{x - 5}$$

Multiply the first rational expression by the reciprocal of the second rational expression.

$$\frac{x^3 + 1}{x^2 - 5x} \cdot \frac{x - 5}{x^2 - 1}$$

Factor each expression if possible and cancel common factors.

$$\frac{(x + 1)(x^2 - x + 1)}{x(x - 5)} \cdot \frac{x - 5}{(x - 1)(x + 1)}$$

$$\frac{\cancel{(x + 1)}(x^2 - x + 1)}{x\cancel{(x - 5)}} \cdot \frac{\cancel{x - 5}}{(x - 1)\cancel{(x + 1)}}$$

$$\frac{x^2 - x + 1}{x(x - 1)} \quad \text{or} \quad \frac{x^2 - x + 1}{x^2 - x}$$

Section 4.2 – Multiplying & dividing rational expressions

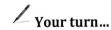 **Your turn...**

Multiply and divide the rational expressions.

1. $\dfrac{x^2+6x}{x^2} \cdot \dfrac{4}{x^2-36}$

2. $\dfrac{k^3-2k^2-8k}{k^2-4} \cdot \dfrac{6k-12}{k^2-9k+20}$

3. $\dfrac{x^2-16}{3x+21} \cdot \dfrac{12x}{5x+20} \cdot \dfrac{x+7}{x^2-4x}$

4. $\dfrac{x^3-81x}{x^2-5x} \div \dfrac{x^2-81}{x^2-3x-10}$

5. $\dfrac{3x^2-14x+8}{x^2-16} \div \dfrac{27x-18}{5x+20}$

6. $\dfrac{10-2x}{6x^2-72x} \div \dfrac{2x^2-50}{6x}$

7. $\dfrac{x^3-1}{x^2-1} \cdot \dfrac{x+1}{2x}$

8. $\dfrac{x^3+8}{x^2-3x} \div \dfrac{x^2-4}{x-3}$

For more practice, access the online homework. See your syllabus for details.

Section 4.3 – Adding & Subtracting Rational Expressions

- Adding & subtracting rational expressions – like denominators
- Adding & subtracting rational expressions – unlike denominators

---

❖ *Adding & subtracting rational expressions – like denominators*

To add or subtract rational expressions with the **same or like denominators**, simply keep the denominator and add the numerators. Factor, if possible, and simplify any common factors from the numerator and denominator.

Example 1

Add the rational expressions.

$$\frac{2}{x} + \frac{1}{x} = \frac{2+1}{x} = \frac{3}{x}$$

Example 2

Subtract the rational expressions.

$$\frac{3x}{x^3} - \frac{11}{x^3}$$

$$\frac{3x-11}{x^3}$$

Example 3

Add and subtract the rational expressions.

$$\frac{x^2}{x+8} + \frac{7x}{x+8} - \frac{8}{x+8}$$

$$\frac{x^2 + 7x - 8}{x+8}$$

Factor the numerator. Cancel common factors.

$$\frac{(x+8)(x-1)}{x+8} = \frac{\cancel{(x+8)}(x-1)}{\cancel{x+8}} = \mathbf{x-1}$$

Example 4

Subtract the rational expressions.

$$\frac{x^2}{x^2 - 7x} - \frac{5x + 14}{x^2 - 7x}$$

$$\frac{x^2 - (5x + 14)}{x^2 - 7x} = \frac{x^2 - 5x - 14}{x^2 - 7x} \qquad \text{(Distribute the negative sign.)}$$

Factor the numerator and denominator. Cancel common factors.

$$\frac{(x - 7)(x + 2)}{x(x - 7)} = \frac{\cancel{(x - 7)}(x + 2)}{x\cancel{(x - 7)}}$$

$$\frac{x + 2}{x}$$

❖ *Adding & subtracting rational expressions – unlike denominators*

The Least Common Denominator (LCD) is defined as the smallest positive number that can be used for all denominators of two or more fractions.

To add or subtract rational expressions with **different or unlike denominators,** first find the least common denominator and <u>make sure that all rational expressions have the same denominator</u>. It is done by multiplying the numerator and denominator by the missing factor. Once they have the same denominator, keep the denominator and add the numerators. Factor, if possible, and simplify any common factors from the numerator and denominator.

Before we start adding or subtracting rational expressions with unlike denominators, it is good to review how to find the least common denominator.

To **find the least common denominator (LCD),** break down the denominators into factors. Choose all the factors, but make sure that if a factor is repeated, choose the highest number of times it is repeated. Let us see how to find the common denominator for the rational expressions below.

Example 5

Find the Least Common Denominator (LCD).

a) $\dfrac{1}{2x}, \dfrac{1}{8}, \dfrac{1}{3}$

Break down each denominator into factors: $2 \cdot x$; $\;2 \cdot 2 \cdot 2$; $\;3$; the **LCD** *is* $2 \cdot 2 \cdot 2 \cdot 3 \cdot x = \mathbf{24x}$.

b) $\dfrac{1}{x}, \dfrac{2}{x^2}, \dfrac{3}{x^3}, \dfrac{1}{5}$

Break down each denominator into factors: x; $\;x \cdot x$; $\;x \cdot x \cdot x$; $\;5$; the **LCD** *is* $x \cdot x \cdot x \cdot 5 = \mathbf{5x^3}$.

c) $\dfrac{1}{x^2-16}, \dfrac{5-x}{x+4}$

Break down each denominator into factors: $(x+4)(x-4)$; $x+4$; the **LCD** *is* $(x+4)(x-4)$.

d) $\dfrac{2}{(x+1)^2}, -\dfrac{3}{x^2+3x+2}, \dfrac{4}{x-7}$

Break down each denominator into factors: $(x+1)(x+1)$; $(x+1)(x+2)$; $x-7$;

$$\text{the } \textbf{LCD} \text{ is } (x+1)^2(x+2)(x-7).$$

Example 6

Subtract the rational expressions.

$$\dfrac{2}{x}-\dfrac{13}{x^2}$$

The LCD is x^2.

$\dfrac{2(x)}{x(x)}-\dfrac{13}{x^2}$ (Put each expression on the same denominator.)

$\dfrac{2(x)}{x^2}-\dfrac{13}{x^2} = \dfrac{2x-13}{x^2}$ (Keep the denominator and add the numerators.)

Example 7

Add the rational expressions.

$$\dfrac{3}{4}+\dfrac{7}{6x}+\dfrac{5}{2x}$$

The LCD is $12x$.

$\dfrac{3(3x)}{4(3x)}+\dfrac{7(2)}{6x(2)}+\dfrac{5(6)}{2x(6)}$ (Put each expression on the same denominator.)

$\dfrac{9x}{12x}+\dfrac{14}{12x}+\dfrac{30}{12x}$ (Keep the denominator and add the numerators.)

$$\dfrac{9x+44}{12x}$$

Example 8

Subtract the rational expressions.

$$\frac{1}{x} - \frac{x}{x^2 - 6x}$$

Factor the denominator.

$$\frac{1}{x} - \frac{x}{x(x - 6)}$$

The LCD is $x(x - 6)$.

$$\frac{1(x - 6)}{x(x - 6)} - \frac{x}{x(x - 6)}$$ (Put each expression on the same denominator.)

$$\frac{1(x - 6) - x}{x(x - 6)}$$ (Keep the denominator and add the numerators.)

$$\frac{x - 6 - x}{x(x - 6)} \rightarrow \frac{-6}{x(x - 6)}$$ (Combine like terms in the numerator.)

Example 9

Add the rational expressions.

$$\frac{2}{x^2 + 6x + 8} + \frac{x}{x + 2} + \frac{1}{x + 4}$$

Factor the denominator.

$$\frac{2}{(x + 2)(x + 4)} + \frac{x}{x + 2} + \frac{1}{x + 4}$$

The LCD is $(x + 2)(x + 4)$.

$$\frac{2}{(x + 2)(x + 4)} + \frac{x(x + 4)}{(x + 2)(x + 4)} + \frac{1(x + 2)}{(x + 4)(x + 2)}$$ (Put each expression on the same denominator.)

$$\frac{2 + x^2 + 4x + x + 2}{(x + 2)(x + 4)}$$ (Keep the denominator and add the numerators.)

Combine like terms, factor, and cancel common factors.

$$\frac{x^2 + 5x + 4}{(x + 2)(x + 4)} = \frac{(x + 1)(x + 4)}{(x + 2)(x + 4)} = \frac{(x + 1)\cancel{(x + 4)}}{(x + 2)\cancel{(x + 4)}} = \frac{(x + 1)}{(x + 2)}$$

Example 10

Subtract the rational expressions.

$$\frac{x}{x^2 - 3x + 2} - \frac{17}{x^2 + 2x - 8}$$

Factor the denominators.

$$\frac{x}{(x-1)(x-2)} - \frac{17}{(x+4)(x-2)}$$

The LCD is $(x - 1)(x - 2)(x + 4)$.

$$\frac{x(x+4)}{(x-1)(x-2)(x+4)} - \frac{17(x-1)}{(x+4)(x-2)(x-1)}$$

(Put each expression on the same denominator.)

$$\frac{x^2 + 4x - 17x + 17}{(x-1)(x-2)(x+4)} = \frac{x^2 - 13x + 17}{(x-1)(x-2)(x+4)}$$

(Keep the denominator and add the numerators.)

Section 4.3 – Adding & subtracting rational expressions

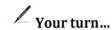 **Your turn...**

Add and subtract the rational expressions.

1. $\dfrac{5}{x} + \dfrac{3}{x}$

2. $\dfrac{2x}{x^3} - \dfrac{15}{x^3}$

3. $\dfrac{x^2}{x-6} + \dfrac{x}{x-6} - \dfrac{42}{x-6}$

4. $\dfrac{x^2}{x^2-7x} - \dfrac{2x+35}{x^2-7x}$

5. $\dfrac{1}{x} - \dfrac{12}{x^2}$

6. $\dfrac{2}{5} + \dfrac{1}{6x} + \dfrac{3}{2x}$

7. $\dfrac{1}{x} - \dfrac{x}{x^2-4x}$

8. $\dfrac{2}{x^2+7x+12} + \dfrac{x}{x+3} + \dfrac{1}{x+4}$

9. $\dfrac{x}{x^2-4x+3} - \dfrac{19}{x^2+x-12}$

For more practice, access the online homework. See your syllabus for details.

Section 4.4 – Simplifying Complex Fractions

❖ Method: clear fractions

❖ *Method: clear fractions*

There are a couple of ways to simplify complex fractions. In this section, we will look at one way, clearing out the fractions. Find the least common denominator using all the denominators of all the fractions and multiply the numerator of every term of the complex fraction by it. Then simplify.

Example 1

Simplify the complex fraction.

$$\frac{\frac{1}{3}+\frac{1}{m}}{\frac{5}{6}-\frac{3}{m}}$$

The LCD is $6m$. Multiply all the numerators by $6m$.

$$\frac{6m(\frac{1}{3}+\frac{1}{m})}{6m(\frac{5}{6}-\frac{3}{m})}$$

$$\frac{\frac{(6m)1}{3}+\frac{(6m)1}{m}}{\frac{(6m)5}{6}-\frac{(6m)3}{m}}=\frac{2m+6}{5m-18}$$

Example 2

Simplify the complex fraction.

$$\frac{1-\frac{25}{n^2}}{\frac{1}{n}+\frac{5}{n^2}}$$

The LCD is n^2. Multiply all the numerators by n^2.

$$\frac{n^2\left(1 - \frac{25}{n^2}\right)}{n^2\left(\frac{1}{n} + \frac{5}{n^2}\right)}$$

$$\frac{1(n^2) - \frac{25(n^2)}{n^2}}{\frac{1(n^2)}{n} + \frac{5(n^2)}{n^2}} = \frac{1(n^2) - \frac{25\,\cancel{(n^2)}}{\cancel{n^2}}}{\frac{1(n^2)}{n} + \frac{5\,\cancel{(n^2)}}{\cancel{n^2}}} = \frac{n^2 - 25}{n + 5}$$

Factor. Cancel common factors.

$$\frac{(n+5)(n-5)}{n+5} = \frac{\cancel{(n+5)}\,(n-5)}{\cancel{n+5}}$$

$$n - 5$$

Example 3

Simplify the complex fraction.

$$\frac{3x + 5}{\frac{1}{x} + \frac{1}{x^2}}$$

The LCD is x^2. Multiply all the numerators by x^2.

$$\frac{x^2(3x + 5)}{x^2\left(\frac{1}{x} + \frac{1}{x^2}\right)}$$

$$\frac{3x(x^2) + 5(x^2)}{\frac{1(x^2)}{x} + \frac{1(x^2)}{x^2}} = \frac{3x(x^2) + 5(x^2)}{\frac{1(x\,\cancel{x})}{\cancel{x}} + \frac{1(\cancel{x^2})}{\cancel{x^2}}}$$

$$\frac{3x^3 + 5x^2}{x + 1}$$

Example 4

Simplify the complex fraction.

$$\frac{\frac{7}{x} + 3}{\frac{7}{x}}$$

The LCD is x. Multiply all the numerators by x.

$$\frac{x\left(\frac{7}{x} + 3\right)}{x\left(\frac{7}{x}\right)}$$

$$\frac{\frac{7(x)}{x} + 3(x)}{\frac{7(x)}{x}} = \frac{\frac{7(\cancel{x})}{\cancel{x}} + 3(x)}{\frac{7(\cancel{x})}{\cancel{x}}} = \frac{7 + 3x}{7}$$

Section 4.4 – Simplifying complex fractions

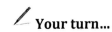 **Your turn...**

Simplify the complex fractions.

1. $\dfrac{\dfrac{1}{2}+\dfrac{1}{m}}{\dfrac{5}{4}-\dfrac{3}{m}}$

2. $\dfrac{1-\dfrac{49}{n^2}}{\dfrac{1}{n}+\dfrac{7}{n^2}}$

3. $\dfrac{1-\dfrac{81}{n^2}}{\dfrac{1}{n}+\dfrac{9}{n^2}}$

4. $\dfrac{2x+3}{\dfrac{1}{x}+\dfrac{9}{x^2}}$

5. $\dfrac{x+5}{\dfrac{1}{x}+\dfrac{5}{x^2}}$

6. $\dfrac{\dfrac{2}{x}+3}{\dfrac{2}{x}}$

For more practice, access the online homework. See your syllabus for details.

Section 4.5 - Solving Rational Equations and Literal Equations

❖ Solving rational equations
❖ Solving literal equations

❖ *Solving rational equations (Clear denominators)*

A rational equation is an equation which contains one or more fractional terms.

To solve rational equations, we use the same method discussed in the previous section with complex fractions. Find the least common denominator using all the denominators and multiply the numerator of each term by it. All the denominators will be cleared if the correct least common denominator is used. Then solve the equation. Once a solution is found, plug it back in the original equation. Remember, there cannot be a zero in the denominator of a rational expression or equation. If the denominator is zero, the answer is **no solution**.

Example 1

Solve.

$$\frac{1}{x} + \frac{2}{3x} = \frac{1}{5}$$

The LCD is $15x$. Multiply all the numerators by $15x$.

$$15x \left(\frac{1}{x} + \frac{2}{3x} = \frac{1}{5} \right)$$

$$\frac{1(15x)}{x} + \frac{2(15x)}{3x} = \frac{1(15x)}{5}$$

$$\frac{1(15\not{x})}{\not{x}} + \frac{2(15\not{x})}{3\not{x}} = \frac{1(15x)}{5} \qquad \text{(Simplify.)}$$

$$15 + 10 = 3x$$

$$25 = 3x \rightarrow x = \frac{25}{3}$$

This answer is a solution. If $x = \frac{25}{3}$ is substituted into the original equation, the denominator is not zero.

Example 2

Solve.

$$1 - \frac{1}{x} = \frac{30}{x^2}$$

The LCD is x^2. Multiply all the numerators by x^2.

$$x^2 \left(1 - \frac{1}{x} = \frac{30}{x^2}\right)$$

$$1(x^2) - \frac{1(x^2)}{x} = \frac{30(x^2)}{x^2} \qquad \text{(Simplify.)}$$

$$1(x^2) - \frac{1(x\cancel{x})}{\cancel{x}} = \frac{30(\cancel{x^2})}{\cancel{x^2}}$$

$$x^2 - x = 30$$

$$x^2 - x - 30 = 0 \qquad \text{(Set it equal to zero by subtracting 30 from both sides.)}$$

$$(x - 6)(x + 5) = 0 \qquad \text{(Factor.)}$$

$$x - 6 = 0 \rightarrow x = 6 \qquad \text{(Solve each factor.)}$$

$$x + 5 = 0 \rightarrow x = -5 \qquad \text{(Solve each factor.)}$$

Both answers are solutions. If $x = -5$ and 6 are substituted into the original equation, the denominator is not zero.

Example 3

Solve.

$$\frac{1}{4} + \frac{1}{7} = \frac{1}{k}$$

The LCD is $28k$. Multiply all the numerators by $28k$.

$$28k \left(\frac{1}{4} + \frac{1}{7} = \frac{1}{k}\right)$$

$$\frac{1(28k)}{4} + \frac{1(28k)}{7} = \frac{1(28k)}{k}$$

$$\frac{1(28k)}{4} + \frac{1(28k)}{7} = \frac{1(28\cancel{k})}{\cancel{k}} \qquad \text{(Simplify.)}$$

$$7k + 4k = 28$$

$$11k = 28 \rightarrow k = \frac{28}{11}$$

This answer is a solution. If $k = \frac{28}{11}$ is substituted into the original equation, the denominator is not zero.

Example 4

Solve.

$$\frac{4x}{x-3} = \frac{4x+1}{x+2}$$

The LCD is $(x-3)(x+2)$. Multiply all the numerators by $(x-3)(x+2)$.

Note that cross-multiplication can also be used in this case.

$$[(x-3)(x+2)](\frac{4x}{x-3} = \frac{4x+1}{x+2})$$

$$\frac{[(x-3)(x+2)]4x}{x-3} = \frac{[(x-3)(x+2)](4x+1)}{x+2}$$

$$\frac{4x\,\cancel{(x-3)}\,(x+2)}{\cancel{x-3}} = \frac{(x-3)\,\cancel{(x+2)}\,(4x+1)}{\cancel{x+2}} \qquad \text{(Simplify.)}$$

$$4x(x+2) = (x-3)(4x+1)$$

$$4x^2 + 8x = 4x^2 + x - 12x - 3 \qquad \text{(Distribute the } 4x \text{ and FOIL.)}$$

$$4x^2 + 8x = 4x^2 - 11x - 3 \qquad \text{(Combine like terms.)}$$

$$8x = -11x - 3 \qquad \text{(Subtract } 4x^2 \text{ from both sides.)}$$

$$8x + 11x = -3 \qquad \text{(Add } 11x \text{ to both sides.)}$$

$$19x = -3 \rightarrow x = -\frac{3}{19}$$

This answer is a solution. If $x = -\frac{3}{19}$ is substituted into the original equation, the denominator is not zero.

Example 5

Solve.

$$\frac{13}{x-1} + \frac{7}{x-1} = 5 + \frac{20}{x-1}$$

The LCD is $(x-1)$. Multiply all the numerators by $(x-1)$.

$$(x-1)\left[\frac{13}{x-1} + \frac{7}{x-1} = 5 + \frac{20}{x-1}\right]$$

$$\frac{13(x-1)}{x-1} + \frac{7(x-1)}{x-1} = 5(x-1) + \frac{20(x-1)}{x-1}$$

$$\frac{13\cancel{(x-1)}}{\cancel{x-1}} + \frac{7\cancel{(x-1)}}{\cancel{x-1}} = 5(x-1) + \frac{20\cancel{(x-1)}}{\cancel{x-1}} \qquad \text{(Simplify.)}$$

$$13 + 7 = 5(x-1) + 20$$

$$20 = 5x - 5 + 20 \qquad \text{(Distribute the 5.)}$$

$$20 = 5x + 15 \qquad \text{(Combine like terms.)}$$

$$20 - 15 = 5x \qquad \text{(Subtract 15 from both sides.)}$$

$$5 = 5x \rightarrow x = 1$$

The answer is **no solution**. If $x = 1$ is substituted into the original equation, the denominator is zero.

Example 6

It takes Alan 3 hours to wax the first floor of building 1. It takes Frank 6 hours to wax the same floor. How long will it take if Alan and Frank work together?

Set up explanations – It takes Alan 3 hours to wax the floor. It means that Alan waxes 1/3 of the floor in one hour. Similarly, it takes Frank 6 hours to wax the floor, then it means that Frank waxes 1/6 of the floor in one hour.

$$\frac{1}{3} + \frac{1}{6} = \frac{1}{t}$$

The LCD is $6t$. Multiply all the numerators by $6t$.

$$6t\left(\frac{1}{3} + \frac{1}{6} = \frac{1}{t}\right)$$

$$\frac{1(6t)}{3} + \frac{1(6t)}{6} = \frac{1(6t)}{t}$$

$$\frac{1(6t)}{3} + \frac{1(6t)}{6} = \frac{1(6\cancel{t})}{\cancel{t}} \qquad \text{(Simplify.)}$$

$$2t + t = 6$$

$$3t = 6 \rightarrow t = \mathbf{2\ hours}$$

It will take 2 hours to wax the floor if they work together.

Example 7

Lisa estimates that it may take her 12 months to work on an algebra book. Megan estimates that it will take her 18 months to work on a similar algebra book. They realize that it may take them fewer months if they work together. Were they right? How long will it take if they work together?

Set up explanations – If it takes take Lisa 12 months to work on the book. It means that Lisa completes 1/12 of the book in one month. Similarly, if it takes Megan 18 months to complete the book, then it means that Megan completes 1/18 of the book in one month.

$$\frac{1}{12} + \frac{1}{18} = \frac{1}{t}$$

The LCD is $36t$. Multiply all the numerators by $36t$.

$$36t\left(\frac{1}{12} + \frac{1}{18} = \frac{1}{t}\right)$$

$$\frac{1(36t)}{12} + \frac{1(36t)}{18} = \frac{1(36t)}{t}$$

$$\frac{1(36t)}{12} + \frac{1(36t)}{18} = \frac{1(36\cancel{t})}{\cancel{t}} \qquad \text{(Simplify.)}$$

$$3t + 2t = 36$$

$$5t = 36 \rightarrow t = \frac{36}{5} \rightarrow t = \mathbf{7.2\ months}$$

They were right. If they work together, it will take approximately 7 months to complete the book.

Example 8

Two Phi Theta Kappa advisors are traveling to a conference in Miami in two separate cars with 4 officers in each car. Advisor Lucas drives 10 mph faster than advisor Chloe. When Lucas has traveled 120 miles, Chloe has traveled 100 miles. What is the rate of each advisor?

For this example, we will use the formula $D = RT$ (D is distance, R is rate, and T is time).

For **Chloe**, we have: For **Lucas**, we have:

$D = RT$ $D = RT$

$100 = RT$ $120 = (R + 10)T$

$\dfrac{100}{R} = T$ (Isolate T.) $\dfrac{120}{R+10} = T$ (Isolate T.)

Nothing is mentioned about the time. We are going to set the times equal to each other in order to solve for the rate or use the substitution method.

$$\dfrac{120}{R+10} = \dfrac{100}{R}$$

The LCD is R(R+10).

$$\dfrac{120\,R(R+10)}{R+10} = \dfrac{100R(R+10)}{R}$$ (Clear the fraction by multiplying the numerators by the LCD.)

Note: Cross multiplication can also be used.

$120R = 100(R + 10)$

$120R = 100R + 1000$ (Distribute the 100.)

$120R - 100R = 1000$ (Subtract 100R from both sides.)

$20R = 1000$ (Combine like terms.)

$R = 50$ Mph   (Divide both sides by 20.)

Chloe is traveling at **50 mph**.

Advisor Lucas drives 10 mph faster than advisor Chloe, Lucas' rate is $R = 50 + 10 = 60$ mph.

Lucas is traveling at **60 mph**.

❖ *Solving literal equations*

A literal equation is a formula used to express mathematical relationships or rule in symbols. When solving literal equations that involve rational expressions, we treat all variables, other than the one we solve for, as constants.

We will specify which variable or letter to solve for when solving literal equations.

Example 9

Given the equation below, **solve for C.**

$$A = \frac{B}{C} + 1$$

Clear the fraction by multiplying every term by C (LCD).

$$A(C) = \frac{B(C)}{C} + 1(c)$$

$AC = B + C$ (Clear the denominator.)

$AC - C = B$ (Subtract C from both sides in order to get all the C's on one side.)

$C(A - 1) = B$ (To isolate C, factor C.)

$\frac{C(A-1)}{(A-1)} = \frac{B}{(A-1)}$ (Divide both sides by $A - 1$.)

$$C = \frac{B}{A-1}$$

Example 10

Given the equation below, **solve for M.**

$$\frac{1}{M} + \frac{1}{N} = P$$

Clear the fraction by multiplying every term by MN (LCD).

$$\frac{1(MN)}{M} + \frac{1(MN)}{N} = P(MN)$$

$N + M = MNP$ (Clear denominator)

$N = MNP - M$ (Subtract M from both sides in order to have all M's on one side.)

$N = M(NP - 1)$ (To isolate M, factor M.)

$\frac{N}{(NP-1)} = \frac{M(NP-1)}{(NP-1)}$ (Divide both sides by $NP - 1$.)

$$\frac{N}{NP-1} = M$$

Example 11

Given the equation below, **solve for R.**

$$F = \frac{MV^2}{R}$$

$$\frac{F}{1} = \frac{MV^2}{R} \qquad \text{(Write F as a fraction.)}$$

$$FR = MV^2 \qquad \text{(Use cross-multiplication.)}$$

$$\boldsymbol{R} = \frac{MV^2}{F} \qquad \text{(Divide both sides by F.)}$$

Example 12

Given the equation below, **solve for K.**

$$E = \frac{KQ}{R^2}$$

$$\frac{E}{1} = \frac{KQ}{R^2} \qquad \text{(Write E as a fraction.)}$$

$$ER^2 = KQ \qquad \text{(Use cross-multiplication.)}$$

$$\frac{ER^2}{Q} = \boldsymbol{K} \qquad \text{(Divide both sides by Q.)}$$

Section 4.5 – Solving rational equations & literal equations

✏ **Your turn...**

Solve the rational equations.

1. $\dfrac{1}{x} + \dfrac{2}{3x} = \dfrac{1}{4}$

2. $1 - \dfrac{1}{x} = \dfrac{42}{x^2}$

3. $\dfrac{1}{3} + \dfrac{1}{7} = \dfrac{1}{k}$

4. $\dfrac{2x}{x-1} = \dfrac{2x+1}{x+3}$

5. $\dfrac{10}{s-1} + \dfrac{2}{s-1} = 3 + \dfrac{12}{s-1}$

6. It takes Hamid 2 hours to wax the first floor of building 1. It takes Juan 3 hours to wax the same floor. How long will it take if Hamid and Juan work together?

7. Zawadi estimates that it may take her 3 months to review an algebra book. Martha estimates that it will take her 5 months to review a similar algebra book. They realize that it may take them fewer months if they work together. How long will it take if they work together?

8. Two Phi Theta Kappa advisors are traveling to a conference in West Palms Beach in two separate cars with 4 officers in each car. Advisor Toby drives 15 mph faster than advisor Luciano. When Toby has traveled 110 miles, Luciano has traveled 85 miles. What is the rate of each advisor?

Solve literal equations.

9. $A = \dfrac{B}{C} + 1$; for B

10. $\dfrac{1}{M} + \dfrac{1}{N} = P$; for N

11. $F = \dfrac{MV^2}{R}$; for M

12. $E = \dfrac{KQ}{R^2}$; for Q

For more practice, access the online homework. See your syllabus for details.

Chapter 4: Review of Terms, Concepts, and Formulas

- To **simplify rational expressions** (expressions in ratio form with a variable in the denominator), factor all the expressions in the numerator and denominator if applicable. Then simplify any common factors from the numerator and denominator. Remember that only factors (multiplication) can be cancelled out.

- To find the **domain** of a rational function, set the denominator equal to zero and solve it. The value(s) found, if any, will not be in the domain. In other words, the domain will be all real numbers except the value(s) found when solving the denominator of the function.

- To **multiply rational expressions**, simplify the rational expressions first by factoring all the expressions. Simplify any common factors from the numerator and denominator. Then multiply the results.

- To **divide rational expressions**, keep the first rational expression as is and multiply it by the reciprocal of the second rational expression. Factor all the expressions. Simplify any common factors from the numerator and denominator. Multiply the results.

- To **add or subtract rational expressions** with the **same or like denominators**, simply keep the denominator and add the numerators. Factor, if possible, and simplify any common factors from the numerator and denominator.

- To **add or** subtract rational expressions with **different or unlike denominators**, first find the least common denominator and <u>make sure that all rational expressions have the same denominator</u>. It is done by multiplying the numerator and denominator by the missing factor. Once they have the same denominator, keep the denominator and add the numerators. Factor, if possible, and simplify any common factors from the numerator and denominator.

- To **find the least common denominator (LCD)**, break down the denominators into factors. Choose all the factors, but make sure that if a factor is repeated, choose the highest number of times it is repeated.

- To **simplify complex fractions**, one way is to find the least common denominator using all the denominators of all the fractions and multiply the numerator of every term of the complex fraction by it. Then simplify.

- To **solve rational equations**, one way is to find the least common denominator using all the denominators and multiply the numerator of each term by it. All the denominators will be cleared if the correct least common denominator is used. Then solve the equation. Once a solution is found, plug it back in the original equation. Remember, there cannot be a zero in the denominator of a rational expression or equation. If the denominator is zero, the answer is **no solution**.

- A **literal equation** is a formula used to express mathematical relationships or rule in symbols. When solving literal equations that involve rational expressions, we treat all variables, other than the one we solve for, as constants. The one to solve for is specified in the instructions.

CHAPTER 5
RADICAL EXPRESSIONS, EQUATIONS, & FUNCTIONS

Section 5.1 – Simplifying Radical Expressions & Evaluating Radical Functions

- ❖ Index two
- ❖ Index three and higher
- ❖ Same index and same power
- ❖ Root approximation
- ❖ Expressions including variables
- ❖ Applications involving radicals
- ❖ Evaluation of radical functions

Before simplifying a radical expression, be familiar with the index and what it means. Let us review:

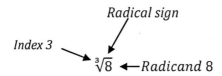

a) $\sqrt{16}$

This is the square root of 16. The index is 2 and it is not shown. The answer to this expression is 4. Sixteen is the result of multiplying the number 4 twice (Index 2).

b) $\sqrt[3]{64}$

This is the cube root of 64. The index is 3 and it is shown in the upper left corner of the radical sign. The answer to this expression is 4. Sixty four is the result of multiplying the number 4 three times (Index 3).

c) $\sqrt[4]{16}$

This is the fourth root of 16. The index is 4 and it is shown in the upper left corner of the radical sign. The answer to this expression is 2. Sixteen is the result of multiplying the number 2 four times (Index 4).

❖ *Index two*

Example 1
Simplify the radical expressions.

Note: The index is two (or even) so the radicand cannot be negative.

a) $\sqrt{49} = 7$ c) $\sqrt{121} = 11$

b) $\sqrt{100} = 10$ d) $\sqrt{36} = 6$

e) $\sqrt{\dfrac{9}{25}} = \dfrac{3}{5}$

f) $-\sqrt{49} = -(7) = -7$

g) $\sqrt{-81}$ = Not a real number $[(-9)(-9) = 81$ and $(9)(9) = 81]$

❖ **Index three and higher**

Example 2

Simplify the radical expressions.

Note: For odd index, the radicand can be negative. For even index, the radicand cannot be negative.

a) $\sqrt[5]{32} = 2$

b) $\sqrt[4]{81} = 3$

c) $\sqrt[3]{-8} = -2$

d) $\sqrt[4]{-256}$ = Not a real number

e) $\sqrt[6]{\dfrac{1}{64}} = \dfrac{1}{2}$

f) $\sqrt[4]{\dfrac{16}{625}} = \dfrac{2}{5}$

g) $-\sqrt[3]{-343} = -(-7) = 7$

❖ **Same index and same power**

What happens when the index is the same as the power? The radical goes away when the expression has the same index and power. The quickest way to simplify such expression is to look at the index.

If the index is even and the expression under the radical is positive, the answer is the positive expression without the radical sign. If the index is even and the expression under the radical is negative, the answer is still a positive expression without the radical sign. $\sqrt{5^2} = 5$; $\sqrt{(-5)^2} = 5$

If the index is odd and the expression under the radical is positive, the answer is the positive expression without the radical sign. If the index is odd and the expression under the radical is negative, the answer is the negative expression without the radical sign. $\sqrt[3]{4^3} = 4$; $\sqrt[3]{(-4)^3} = -4$

Example 3

Simplify the radical expressions.

a) $\sqrt[5]{4^5} = 4$

b) $\sqrt{3^2} = 3$

c) $\sqrt{(-16)^2} = 16$

d) $\sqrt[3]{5^3} = 5$

e) $\sqrt[3]{(-6)^3} = -6$

f) $\sqrt[4]{8^4} = 8$

g) $\sqrt[4]{(-7)^4} = 7$

h) $\sqrt[6]{(-3)^6} = 3$

i) $\sqrt{x^2} = x$ (x is positive.)

j) $\sqrt[3]{(x+1)^3} = x + 1$

❖ *Root approximation*

When simplifying numbers that are not perfect squares, cubes, fourths, and fifths, we can use a calculator to find decimal approximations of them. We express the answer using this approximation symbol ($\approx$).

Example 4

Use a calculator to simplify the following expressions and round your final answers to three decimal places.

a) $\sqrt{10} \approx 3.162$

b) $\sqrt[3]{7} \approx 1.913$

c) $\sqrt[4]{20} \approx 2.115$

d) $\sqrt[5]{14} \approx 1.695$

❖ *Expressions including variables*

When simplifying a radical expression, look at the index. If the index is 2, the number must be broken up into 2 factors where one of the factors is a perfect square. If the index is 3, the number must be broken up into 2 factors where one of the factors is a perfect cube. If the index is 4, the number must be broken up into 2 factors where one of the factors is a perfect fourth and so on. Then the perfect square or cube or fourth will be without the radical sign and the other factor will have the radical sign. $[\sqrt{20} = \sqrt{4 \cdot 5} = \sqrt{4} \cdot \sqrt{5} = 2\sqrt{5}]$

Example 5

Simplify the radical expressions.

a) $\sqrt{98} = \sqrt{49 \cdot 2} = \sqrt{49} \cdot \sqrt{2} = 7\sqrt{2}$

b) $\sqrt[3]{-54} = \sqrt[3]{-27 \cdot 2} = \sqrt[3]{-27} \cdot \sqrt[3]{2} = -3\sqrt[3]{2}$

c) $\sqrt{4\,x^3y^5} = \sqrt{4\,x^2xy^4y} = \sqrt{4x^2y^4} \cdot \sqrt{xy} = 2xy^2\sqrt{xy}$

d) $\sqrt{81\,l^2m^3n^7} = \sqrt{81\,l^2m^2mn^6n} = \sqrt{81l^2m^2n^6} \cdot \sqrt{mn} = 9lmn^3\sqrt{mn}$

e) $\sqrt[3]{-8a^6b^{10}c^{14}} = \sqrt[3]{-8a^6b^9bc^{12}c^2} = \sqrt[3]{-8a^6b^9c^{12}} \cdot \sqrt[3]{bc^2} = -2a^2b^3c^4\sqrt[3]{bc^2}$

f) $\sqrt[4]{32\,u^3v^4w^5y^7} = \sqrt[4]{16 \cdot 2\,u^3v^4w^4wy^4y^3} = \sqrt[4]{16v^4w^4y^4} \cdot \sqrt[4]{2u^3wy^3} = 2vwy\sqrt[4]{2u^3wy^3}$

g) $\sqrt[5]{32\,u^3v^5w^5y^7} = \sqrt[5]{32 \cdot u^3v^5w^5y^5y^2} = \sqrt[5]{32v^5w^5y^5} \cdot \sqrt[5]{u^3y^2} = 2vwy\sqrt[5]{u^3y^2}$

h) $\sqrt[6]{128\,v^4w^5y^7} = \sqrt[6]{64 \cdot 2v^4w^5y^6y^1} = \sqrt[6]{64y^6} \cdot \sqrt[6]{2v^4w^5y^1} = 2y\sqrt[6]{2v^4w^5y}$

❖ *Applications involving radicals*

Example 6

In order to clean a twenty-foot high window, one of our Osceola maintenance persons, Lorenzo, needs to put a ladder eight feet away from the building. How long of a ladder is needed for this job?

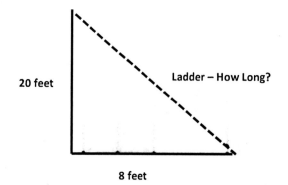

20 feet

Ladder – How Long?

8 feet

This scenario is a right triangle one. Use the Pythagorean Theorem to solve the problem for c or the length of the ladder.

$$c^2 = a^2 + b^2$$

$c^2 = 20^2 + 8^2$ (Replace a and b.)

$c^2 = 464$ (Add.)

$c = \pm\sqrt{464}$ (To remove the square, take the square root of both sides.)

$c = 21.5\,ft$ (Measurement is positive.)

The ladder needs to be 21.5 or approximately **22 ft**.

Example 7

On Denn John Lane, there is a wall with a sign that has the college's old name on it. The goal is to update the sign that is thirty feet tall with the college's new name. The maintenance person, Marco, has a ladder that is forty feet long. How far away from the wall does the ladder need to be in order to reach the sign?

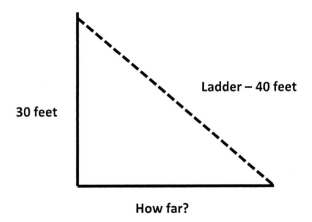

30 feet

Ladder – 40 feet

How far?

This scenario is a right triangle scenario. Use the Pythagorean Theorem to solve the problem for b.

$$c^2 = a^2 + b^2$$

$40^2 = 30^2 + b^2$ (Replace a and c.)

$1600 = 900 + b^2$ (Square the numbers.)

$1600 - 900 = b^2$ (Subtract 900 from both sides.)

$700 = b^2$ (Combine like terms.)

$\pm\sqrt{700} = b$ (To remove the square, take the square root of both sides.)

$b = 26.5\,ft$ (Measurement is positive.)

The ladder needs to be **approximately 27 ft.** away from the wall.

❖ *Evaluation of radical functions*

Evaluating functions mean finding the y −value given an x −value and vice versa. Simply replace the given value in the function.

Example 8

Evaluate the radical functions.

$$f(x) = \sqrt{x + 2} \qquad\qquad g(x) = \sqrt[3]{x - 4} \qquad\qquad h(x) = \sqrt[4]{x - 1}$$

Find $f(-1)$, $f(0)$, $f(2)$, $g(12)$, $g(3)$, $g(6)$, $h(1)$, $h(7)$ and $h(82)$.

a) $f(-1) = \sqrt{-1 + 2} = \sqrt{1} = 1$ (Replace x with −1 in the f function.)

b) $f(0) = \sqrt{0 + 2} = \sqrt{2}$ (Replace x with 0 in the f function.)

c) $f(2) = \sqrt{2 + 2} = \sqrt{4} = 2$ (Replace x with 2 in the f function.)

d) $g(12) = \sqrt[3]{12 - 4} = \sqrt[3]{8} = 2$ (Replace x with 12 in the g function.)

e) $g(3) = \sqrt[3]{3 - 4} = \sqrt[3]{-1} = -1$ (Replace x with 3 in the g function.)

f) $g(6) = \sqrt[3]{6 - 4} = \sqrt[3]{2}$ (Replace x with 6 in the g function.)

g) $h(1) = \sqrt[4]{1 - 1} = \sqrt[4]{0} = 0$ (Replace x with 1 in the h function.)

h) $h(7) = \sqrt[4]{7 - 1} = \sqrt[4]{6}$ (Replace x with 7 in the h function.)

i) $h(82) = \sqrt[4]{82 - 1} = \sqrt[4]{81} = 3$ (Replace x with 82 in the h function.)

Section 5.1 – Simplifying radical expressions & evaluating radical functions

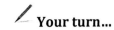 **Your turn...**

Simplify the radical expressions.

1. $\sqrt{25}$

2. $\sqrt{81}$

3. $\sqrt{144}$

4. $\sqrt{\dfrac{4}{81}}$

5. $-\sqrt{25}$

6. $\sqrt{-4}$

7. $\sqrt[3]{8}$

8. $\sqrt[4]{16}$

9. $\sqrt[3]{-27}$

10. $\sqrt[4]{-625}$

11. $\sqrt[3]{\dfrac{1}{27}}$

12. $\sqrt[4]{\dfrac{16}{81}}$

13. $-\sqrt[3]{-125}$

14. $\sqrt{5^2}$

15. $\sqrt{(-12)^2}$

16. $\sqrt[3]{6^3}$

17. $\sqrt[3]{(-4)^3}$

18. $\sqrt[4]{2^4}$

19. $\sqrt[4]{(-2)^4}$

20. $\sqrt{y^2}$

21. $\sqrt[3]{(x+2)^3}$

22. $\sqrt{50}$

23. $\sqrt[3]{-16}$

24. $\sqrt{49\ x^3 y^7}$

25. $\sqrt{100\ l^2 m^4 n^7}$

26. $\sqrt[3]{-27 a^6 b^{12} c^{16}}$

27. $\sqrt[4]{81\ u^3 v^4 w^6 y^9}$

Solve applications involving radicals.

28. In order to clean a twenty- five foot high window, one of our Osceola maintenance persons, Young Soo needs to put a ladder ten feet away from the building. How long of a ladder is needed for this job?

29. On Denn John Lane, there is a wall with a sign that has the college's old name on it. The goal is to update the sign that is twenty feet tall with the college's new name. The maintenance person, Kong, has a ladder that is thirty feet long. How far away from the wall does the ladder need to be in order to reach the sign?

30. Use a calculator to simplify the following expressions and round your final answers to three decimal places.

   a) $\sqrt{35}$

   b) $\sqrt[3]{17}$

   c) $\sqrt[4]{19}$

   d) $\sqrt[5]{11}$

31. Evaluate the radical functions.

$$f(x) = \sqrt{x+5} \qquad g(x) = \sqrt[3]{x+2} \qquad h(x) = \sqrt[4]{x-3}$$

Find $f(-1)$, $f(0)$, $f(4)$, $g(3)$, $g(-10)$, $g(25)$, $h(19)$, $h(5)$ and $h(4)$.

For more practice, access the online homework. See your syllabus for details.

Section 5.2 – Working with Scientific, Standard Notation, and Rational Exponents

- ❖ Converting from standard notation to scientific notation
- ❖ Converting from scientific notation to standard notation
- ❖ Performing arithmetic operations with scientific notation
- ❖ Simplifying rational exponents
- ❖ Performing arithmetic operations with rational exponents

In this section, we discuss how to convert numbers from scientific notation to standard notation and vice versa. These skills are needed when dealing with large numbers or extremely small numbers common in science and engineering. It is easier to write 7.347×10^{22} kg than $73,470,000,000,000,000,000,000$.

A number is in **scientific notation** if it is expressed in the format $n \times 10^p$. To convert a number from standard to **scientific** notation, move the decimal point either to the left or to the right until the number n is between 1 and 10 (1 digit before the decimal point). The power or exponent is p. The sign of p is positive if the decimal point is moved to the left, and it is negative if moved to the right. The value of p is the number of times the decimal point is moved.

To convert a number from scientific notation to **standard**, move the decimal point to the right if p is positive. Move the decimal point to the left if p is negative.

❖ *Converting from standard notation to scientific notation*

Example 1

Convert to scientific notation.

a) $0.0000375 = 3.75 \times 10^{-5}$ (Move right.)

b) $293,000,000 = 2.93 \times 10^{8}$ (Move left.)

For whole numbers, it is understood that the decimal point is at the end of $293,000,00\mathbf{0.}$

❖ *Converting from scientific notation to standard notation*

Example 2

Convert to standard notation.

a) $1.25 \times 10^{4} = 12,500$ (Move right.)

b) $7.6 \times 10^{-3} = 0.0076$ (Move left.)

❖ *Performing arithmetic operations with scientific notation*

<u>Adding and Subtracting</u>

In order to add/subtract numbers in scientific notation, the **exponents must be the same**. If the exponents are the same, keep the exponents and add/subtract the numbers. If the exponents are not the same, make them the same, then keep the exponents and add/subtract the numbers. To make the exponents the same, move the decimal point to the left to increase the exponent or to the right to decrease the exponent.

<u>Multiplying</u>

In order to multiply numbers in scientific notation, the exponents do not need to be the same. Simply multiply the numbers, and **add** the exponents. Make sure the final answer is in scientific notation. (Number between 1 and 10)

<u>Dividing</u>

In order to divide numbers in scientific notation, the exponents do not need to be the same. Simply divide the numbers, and **subtract** the exponents. Make sure the final answer is in scientific notation. (Number between 1 and 10)

Recall: A number is in **scientific notation** if it is expressed in the format $n \times 10^p$. The number n has to be between 1 and 10 (1 digit before the decimal point).

Example 3

Perform the indicated operation.

a) $(2.3 \times 10^3) + (1.65 \times 10^4) = \left(.23 \times \mathbf{10^4}\right) + (1.65 \times 10^4) = 1.88 \times 10^4$

b) $\left(5.7 \times 10^3\right) - \left(4.9 \times 10^2\right) = \left(5.7 \times 10^3\right) - \left(.49 \times \mathbf{10^3}\right) = 5.21 \times 10^3$

c) $\left(6 \times 10^3\right)(3 \times 10^5) = (6 \times 3) \times 10^{3+5} = 18 \times 10^8 = 1.8 \times 10^9$

d) $\dfrac{15 \times 10^6}{6 \times 10^4} = \dfrac{15}{6} \times 10^{6-4} = 2.5 \times 10^2$

e) $(15 \times 10^{-3}) + (18 \times 10^{-4}) = (15 \times 10^{-3}) + \left(\mathbf{1.8} \times \mathbf{10^{-3}}\right) = 16.8 \times 10^{-3} = 1.68 \times 10^{-2}$

f) $(19 \times 10^{-2}) \times (21 \times 10^{-5}) = (399 \times 10^{-2-5}) = 399 \times 10^{-7} = 3.99 \times 10^{-5}$

❖ *Simplifying rational exponents*

An expression with a rational exponent has a base and an exponent in the form of a fraction or ratio. An example of an expression with a rational exponent is $b^{\frac{n}{m}}$ or $4^{\frac{1}{2}}$. To simplify expressions with rational exponents, convert the expression to a radical expression. Looking at the fraction exponent, the denominator of the fraction becomes the index and the numerator of the fraction becomes the power. Then evaluate the radical expression. An example is $4^{\frac{1}{2}} = \sqrt{4} = 2$.

If the rational exponent is negative, before converting to a radical expression, make it positive. An example is $4^{-\frac{1}{2}}$. Take the reciprocal of the expression or write it as 1 over the expression $\left(\frac{1}{4^{\frac{1}{2}}}\right)$. The power becomes positive. Convert to radical as mentioned earlier and evaluate the radical expression. An example is $4^{-\frac{1}{2}} = \frac{1}{4^{\frac{1}{2}}} = \frac{1}{\sqrt{4}} = \frac{1}{2}$.

Example 4

Evaluate expressions with rational exponents.

a) $16^{\frac{1}{2}} = \sqrt{16} = 4$

b) $\left(\frac{4}{9}\right)^{\frac{1}{2}} = \sqrt{\frac{4}{9}} = \frac{2}{3}$

c) $8^{\frac{1}{3}} = \sqrt[3]{8} = 2$

d) $-4^{\frac{1}{2}} = -\sqrt{4} = -2$

e) $(-81)^{\frac{1}{2}} = \sqrt{-81} = $ Not a real number

f) $81^{\frac{1}{4}} = \sqrt[4]{81} = 3$

g) $25^{-\frac{1}{2}} = \frac{1}{25^{\frac{1}{2}}} = \frac{1}{\sqrt{25}} = \frac{1}{5}$

h) $27^{-\frac{1}{3}} = \frac{1}{27^{\frac{1}{3}}} = \frac{1}{\sqrt[3]{27}} = \frac{1}{3}$

i) $(-8)^{-\frac{1}{3}} = \frac{1}{(-8)^{\frac{1}{3}}} = \frac{1}{\sqrt[3]{-8}} = \frac{1}{-2}$ or $-\frac{1}{2}$

Example 5

Evaluate expressions with rational exponents.

a) $64^{\frac{2}{3}} = \left(\sqrt[3]{64}\right)^2 = (4)^2 = 16$

b) $\left(\frac{27}{125}\right)^{\frac{2}{3}} = \left(\sqrt[3]{\frac{27}{125}}\right)^2 = \left(\frac{3}{5}\right)^2 = \frac{9}{25}$

c) $49^{\frac{3}{2}} = \left(\sqrt{49}\right)^3 = (7)^3 = 343$

d) $(-32)^{\frac{3}{5}} = \left(\sqrt[5]{-32}\right)^3 = (-2)^3 = -8$

e) $64^{\frac{2}{6}} = \left(\sqrt[6]{64}\right)^2 = (2)^2 = 4$

f) $81^{\frac{3}{4}} = \left(\sqrt[4]{81}\right)^3 = (3)^3 = 27$

g) $(-8)^{\frac{4}{3}} = \left(\sqrt[3]{-8}\right)^4 = (-2)^4 = 16$

h) $27^{-\frac{2}{3}} = \frac{1}{27^{\frac{2}{3}}} = \frac{1}{(\sqrt[3]{27})^2} = \frac{1}{3^2} = \frac{1}{9}$

i) $-16^{-\frac{3}{4}} = \frac{1}{-16^{\frac{3}{4}}} = \frac{1}{-(\sqrt[4]{16})^3} = \frac{1}{-(2)^3} = \frac{1}{-8} = -\frac{1}{8}$

j) $(-25)^{-\frac{3}{2}} = \frac{1}{(-25)^{\frac{3}{2}}} = \frac{1}{(\sqrt{-25})^3} = $ Not a real number

❖ *Performing arithmetic operations with rational exponents*

When performing arithmetic operations with rational exponents, we can use the same rules that are used for integer exponents. Below is a review of the rules and we will let m and n be rational numbers.

Rules of exponents	
Negative exponent Rule	$x^{-n} = \dfrac{1}{x^n}$
Product Rule	$x^m \cdot x^n = x^{m+n}$
Quotient Rule	$\dfrac{x^m}{x^n} = x^{m-n}, \qquad x \neq 0$
Power Rule	$(x^m)^n = x^{m \cdot n}; \ (x)^0 \text{or} \left(\dfrac{x}{y}\right)^0 = 1; \ x, y \neq 0$
Power Rule (Products)	$(xy)^n = x^n \cdot y^n$
Power Rule (Quotients)	$\left(\dfrac{x}{y}\right)^n = \dfrac{x^n}{y^n}; \ \left(\dfrac{x}{y}\right)^{-n} = \left(\dfrac{y}{x}\right)^n, y \neq 0$

Example 6

Simplify.

a) $\left(x^{\frac{1}{2}}\right)^{\frac{1}{3}} = x^{1/6}$

b) $x^{1/4} \cdot x^{7/4} = x^{\left(\frac{1}{4}+\frac{7}{4}\right)} = x^{\frac{8}{4}} = x^2$

c) $x^{1/5} \cdot x^{1/3} = x^{\left(\frac{1}{5}+\frac{1}{3}\right)} = x^{\left(\frac{3}{15}+\frac{5}{15}\right)} = x^{\frac{8}{15}}$

d) $\dfrac{x^{1/4}}{x^{3/4}} = x^{\left(\frac{1}{4}-\frac{3}{4}\right)} = x^{-2/4} = x^{\frac{-1}{2}} = \dfrac{1}{x^{1/2}}$

e) $\sqrt[3]{x} \cdot \sqrt[4]{x} = x^{\frac{1}{3}} \cdot x^{\frac{1}{4}} = x^{\left(\frac{4}{12}+\frac{3}{12}\right)} = x^{\frac{7}{12}}$

f) $\dfrac{\sqrt[3]{5}}{\sqrt{5}} = \dfrac{5^{\frac{1}{3}}}{5^{\frac{1}{2}}} = 5^{\left(\frac{1}{3}-\frac{1}{2}\right)} = 5^{\left(\frac{2}{6}-\frac{3}{6}\right)} = 5^{\frac{-1}{6}} = \dfrac{1}{5^{1/6}}$

Example 7

Multiply.

Distributive property and FOIL method are used.

a) $x^{\frac{1}{2}}\left(x^{\frac{3}{2}} + x^4\right) = x^{\left(\frac{1}{2}+\frac{3}{2}\right)} + x^{\left(\frac{1}{2}+4\right)} = x^2 + x^{\left(\frac{1}{2}+\frac{8}{2}\right)} = x^2 + x^{9/2}$

b) $x^{\frac{1}{3}}\left(x^{\frac{4}{3}} - 2x + x^5\right) = x^{\left(\frac{1}{3}+\frac{4}{3}\right)} - 2x^{\left(\frac{1}{3}+1\right)} + x^{\left(\frac{1}{3}+5\right)} = x^{\frac{5}{3}} - 2x^{\frac{4}{3}} + x^{16/3}$

c) $\left(x^{\frac{4}{5}} - 2\right)\left(x^{\frac{4}{5}} - 1\right) = x^{\left(\frac{4}{5}+\frac{4}{5}\right)} - x^{\frac{4}{5}} - 2x^{\frac{4}{5}} + 2 = x^{\frac{8}{5}} - 3x^{\frac{4}{5}} + 2$

Example 8

Factor the following expressions with rational exponents.

When factoring expressions with exponents, factor the lowest exponent.

a) $2x^{\frac{-1}{3}} + 5x^{\frac{4}{3}}$

$2\,x^{\frac{-1}{3}} + 5\,x^{\frac{-1}{3}} \cdot x^{\frac{5}{3}}$ (Rewrite both terms using the common factor or lowest exponent.)

$x^{\frac{-1}{3}}\left(2 + 5x^{\frac{5}{3}}\right)$ (Factor the common factor.)

b) $x^{\frac{2}{9}} - x^{\frac{7}{9}}$

$x^{\frac{2}{9}} - x^{\frac{2}{9}} \cdot x^{\frac{5}{9}}$ (Rewrite both terms using the common factor or lowest exponent.)

$x^{\frac{2}{9}}\left(1 - x^{\frac{5}{9}}\right)$ (Factor the common factor.)

c) $\dfrac{1}{2}x^{\frac{-2}{5}} + 3x^{\frac{3}{5}} - \dfrac{1}{7}x^{\frac{-1}{5}}$

$\dfrac{1}{2}\,x^{\frac{-2}{5}} + 3\,x^{\frac{-2}{5}} \cdot x^{\frac{5}{5}} - \dfrac{1}{7}\,x^{\frac{-2}{5}} \cdot x^{\frac{1}{5}}$ (Rewrite all terms using the common factor or lowest exponent.)

$x^{\frac{-2}{5}}\left(\dfrac{1}{2} + 3x - \dfrac{1}{7}x^{\frac{1}{5}}\right)$ (Factor the common factor.)

Section 5.2 – Working with scientific, standard notation, and rational exponents

Your turn...

1. Convert to scientific notation.
    a) 0.0000134
    b) 75,000,000

2. Convert to standard notation.
    a) 2.35×10^6
    b) 4.39×10^{-4}

3. Evaluate expressions with rational exponents.
    a) $25^{\frac{1}{2}}$
    b) $\left(\frac{9}{16}\right)^{\frac{1}{2}}$
    c) $27^{\frac{1}{3}}$
    d) $-16^{\frac{1}{2}}$
    e) $(-64)^{\frac{1}{2}}$
    f) $16^{\frac{1}{4}}$
    g) $49^{-\frac{1}{2}}$
    h) $8^{-\frac{1}{3}}$
    i) $(-8)^{-\frac{1}{3}}$

4. Evaluate expressions with rational exponents.
    a) $27^{\frac{2}{3}}$
    b) $\left(\frac{8}{27}\right)^{\frac{2}{3}}$
    c) $36^{\frac{3}{2}}$
    d) $16^{\frac{3}{4}}$
    e) $(-27)^{\frac{4}{3}}$
    f) $64^{-\frac{2}{3}}$
    g) $-81^{-\frac{3}{4}}$
    h) $(-100)^{-\frac{3}{2}}$

5. Perform the indicated operation.

a) $(3.2 \times 10^3) + (1.4 \times 10^4)$

b) $(1.8 \times 10^2) + (3.51 \times 10^3)$

c) $(2.3 \times 10^2) - (1.65 \times 10^1)$

d) $(7.4 \times 10^{-4}) - (3.6 \times 10^{-5})$

e) $(4.1 \times 10^7)(3.0 \times 10^5)$

f) $(5.0 \times 10^4)(2.5 \times 10^{-3})$

g) $\dfrac{8.7 \times 10^3}{1.4 \times 10^4}$

h) $\dfrac{5.2 \times 10^{-5}}{1.3 \times 10^5}$

6. Perform the indicated operation.

a) $\left(x^{\frac{1}{5}}\right)^{\frac{1}{4}}$

b) $x^{1/5} \cdot x^{9/5}$

c) $x^{2/7} \cdot x^{1/3}$

d) $\dfrac{x^{1/2}}{x^{5/2}}$

e) $\sqrt[5]{x} \cdot \sqrt[6]{x}$

f) $\dfrac{\sqrt{11}}{\sqrt[3]{11}}$

g) $x^{\frac{1}{6}}\left(x^{\frac{5}{6}} + x^2\right)$

h) $x^{\frac{1}{10}}\left(x^{\frac{3}{10}} - 7x + 4x^3\right)$

i) $\left(x^{\frac{1}{2}} + 5\right)\left(x^{\frac{1}{2}} + 7\right)$

7. Factor

a) $4x^{\frac{-1}{2}} + 11x^{\frac{7}{2}}$

b) $x^{\frac{1}{3}} - x^{\frac{7}{3}}$

c) $\frac{1}{4}x^{\frac{-3}{5}} - 9x^{\frac{3}{5}} + \frac{1}{10}x^{\frac{-1}{5}}$

For more practice, access the online homework. See your syllabus for details.

Section 5.3 – Adding & Subtracting Radical Expressions

- ❖ Adding radical expressions
- ❖ Subtracting radical expressions

To add and subtract radical expressions, they must be like radicals. In other words, the expression or number under the radical sign, the radicand, must be the same. They must also have the same index. In this case, just add the coefficients. If the expressions are not like radicals, simplify first, and then add or subtract if possible.

❖ *Adding radical expressions*

Example 1

Add the radical expressions.

a) $\sqrt{7} + 2\sqrt{7} + 9\sqrt{7} = 12\sqrt{7}$

b) $\frac{1}{2}\sqrt{d} + \frac{3}{4}\sqrt{d} + \frac{3}{5}\sqrt{d} = \frac{10}{20}\sqrt{d} + \frac{15}{20}\sqrt{d} + \frac{12}{20}\sqrt{d} = \frac{37}{20}\sqrt{d}$ (Find LCD.)

c) $4\sqrt[3]{11x} + 3\sqrt[3]{11x} + 24\sqrt[3]{11x} = 31\sqrt[3]{11x}$

d) $\sqrt[5]{3y} + 3\sqrt[5]{3y} + 12\sqrt[5]{3y} = 16\sqrt[5]{3y}$

Example 2

Add the radical expressions.

a) $\sqrt{12} + \sqrt{27} + \sqrt{48} = \sqrt{4\cdot 3} + \sqrt{9\cdot 3} + \sqrt{16\cdot 3} = 2\sqrt{3} + 3\sqrt{3} + 4\sqrt{3} = 9\sqrt{3}$

b) $\sqrt{50x^7} + x\sqrt{32x^5} = \sqrt{2\cdot 25\, x^6 x} + x\sqrt{2\cdot 16\, x^4 x}$

$= 5x^3\sqrt{2x} + x\cdot 4\cdot x^2\sqrt{2x}$

$= 5x^3\sqrt{2x} + 4x^3\sqrt{2x} = 9x^3\sqrt{2x}$

c) $\sqrt[3]{8p^4} + \sqrt[3]{125p} = \sqrt[3]{8p^3 p} + \sqrt[3]{125p} = 2p\sqrt[3]{p} + 5\sqrt[3]{p} = (2p+5)\sqrt[3]{p}$

d) $\sqrt{50} + \sqrt{15} = \sqrt{25 \cdot 2} + \sqrt{15} = 5\sqrt{2} + \sqrt{15}$ (Cannot be combined)

e) $\sqrt[4]{32} + \sqrt[4]{162} + \sqrt[4]{\frac{2}{81}} = \sqrt[4]{16 \cdot 2} + \sqrt[4]{81 \cdot 2} + \sqrt[4]{\frac{1}{81} \cdot 2} = 2\sqrt[4]{2} + 3\sqrt[4]{2} + \frac{1}{3}\sqrt[4]{2} = \frac{16}{3}\sqrt[4]{2}$

f) $\sqrt[5]{32} + \sqrt[5]{96} = \sqrt[5]{32} + \sqrt[5]{32 \cdot 3} = 2 + 2\sqrt[5]{3}$ (Cannot be combined)

❖ **Subtracting radical expressions**

Example 3

Subtract the radical expressions.

a) $5\sqrt{kp} - 7\sqrt{kp} - 4\sqrt{kp} = -6\sqrt{kp}$

b) $\sqrt[4]{3} - 5\sqrt[4]{3} = -4\sqrt[4]{3}$

c) $-2\sqrt[3]{5} - 9\sqrt[3]{5} - 13\sqrt[3]{5} = -24\sqrt[3]{5}$

d) $-\sqrt[6]{x} - 5\sqrt[6]{x} - 3\sqrt[6]{x} = -9\sqrt[6]{x}$

Example 4

Subtract the radical expressions.

a) $\sqrt{K^3} - 9\sqrt{K^3} - 2\sqrt{K^3} = -10\sqrt{K^3} = -10\sqrt{k^2 k} = -10k\sqrt{k}$

b) $\sqrt[4]{32} - \sqrt[4]{162} - \sqrt[4]{\frac{2}{81}} = \sqrt[4]{16 \cdot 2} - \sqrt[4]{81 \cdot 2} - \sqrt[4]{\frac{1}{81} \cdot 2} = 2\sqrt[4]{2} - 3\sqrt[4]{2} - \frac{1}{3}\sqrt[4]{2} = -\frac{4}{3}\sqrt[4]{2}$

c) $\sqrt{72} - \sqrt{13} = \sqrt{36 \cdot 2} - \sqrt{13} = 6\sqrt{2} - \sqrt{13}$ (Cannot be combined)

Section 5.3 – Adding & subtracting radical expressions

 **Your turn...**

Add and subtract the radical expressions.

1. $\sqrt{5} + 2\sqrt{5} + 3\sqrt{5}$

2. $\frac{1}{3}\sqrt{a} + \frac{3}{4}\sqrt{a} + \frac{3}{5}\sqrt{a}$

3. $\sqrt{8} + \sqrt{18} + \sqrt{32}$

4. $\sqrt{75x^9} + x\sqrt{48x^7}$

5. $\sqrt[3]{64p^4} + \sqrt[3]{27p}$

6. $\sqrt{75} + \sqrt{15}$

7. $\sqrt[4]{48} - \sqrt[4]{243} - \sqrt[4]{\frac{3}{81}}$

8. $3\sqrt{np} - 5\sqrt{np} - 2\sqrt{np}$

9. $\sqrt{K^5} - 7\sqrt{K^5} - 5\sqrt{K^5}$

10. $2\sqrt[6]{y^7} - 3\sqrt[6]{y^7} - 11\sqrt[6]{y^7}$

For more practice, access the online homework. See your syllabus for details.

Section 5.4 - Multiplying and Dividing Radical Expressions

❖ Multiplying radical expressions – same index
❖ Dividing radical expressions – same index

❖ *Multiplying radical expressions - same index*

In this section, we multiply radical expressions with the same index. To proceed, multiply the numbers or expressions outside the radical together and multiply the numbers or expressions under the radical sign or radicand together. Then if the result needs to be simplified, do so. In the multiplication process, we use the distributive property and the FOIL method when applicable.

Example 1

Multiply the radical expressions.

a) $\sqrt{2} \cdot \sqrt{8} = \sqrt{16} = 4$

b) $\sqrt{3} \cdot \sqrt{6} = \sqrt{18} = \sqrt{9 \cdot 2} = 3\sqrt{2}$

c) $\sqrt[3]{2} \cdot \sqrt[3]{16} \cdot \sqrt[3]{4} = \sqrt[3]{128} = \sqrt[3]{64 \cdot 2} = 4\sqrt[3]{2}$

d) $\left(2\sqrt{x}\right)\left(3\sqrt{2x}\right) = 6\sqrt{2x^2} = 6x\sqrt{2}$

e) $\left(\frac{1}{2}\sqrt{3}\right)\left(\frac{3}{4}\sqrt{8}\right) = \frac{3}{8}\sqrt{24} = \frac{3}{8}\sqrt{4 \cdot 6} = \frac{3}{8} \cdot 2\sqrt{6} = \frac{6}{8}\sqrt{6}$ or $\frac{3}{4}\sqrt{6}$

Example 2

Multiply the radical expressions.

a) $\sqrt{7}\left(2 + \sqrt{3}\right) = 2\sqrt{7} + \sqrt{21}$

b) $\sqrt{2}\left(3\sqrt{8} - \sqrt{10}\right) = 3\sqrt{16} - \sqrt{20} = 3(4) - \sqrt{4 \cdot 5} = 12 - 2\sqrt{5}$

c) $\sqrt{5}\left(\sqrt{10} + \sqrt{8}\right) = \sqrt{50} + \sqrt{40} = \sqrt{25 \cdot 2} + \sqrt{4 \cdot 10} = 5\sqrt{2} + 2\sqrt{10}$

Example 3

Multiply the radical expressions.

a) $(\sqrt{3} + \sqrt{5})(\sqrt{6} + \sqrt{3}) = \sqrt{18} + \sqrt{9} + \sqrt{30} + \sqrt{15} = \sqrt{9 \cdot 2} + 3 + \sqrt{30} + \sqrt{15} =$
   $3\sqrt{2} + 3 + \sqrt{30} + \sqrt{15}$

b) $(1 + \sqrt{7})(\sqrt{5} - 4) = \sqrt{5} - 4 + \sqrt{35} - 4\sqrt{7}$

c) $(\sqrt{2} + 3)(\sqrt{2} - 3) = \sqrt{4} - 3\sqrt{2} + 3\sqrt{2} - 9 = 2 - 9 = -7$

d) $(\sqrt{6} + 7)(\sqrt{6} + 1) = \sqrt{36} + \sqrt{6} + 7\sqrt{6} + 7 = 6 + 8\sqrt{6} + 7 = 13 + 8\sqrt{6}$

e) $(\sqrt{x} + 7)(\sqrt{x} + 1) = \sqrt{x^2} + \sqrt{x} + 7\sqrt{x} + 7 = x + 8\sqrt{x} + 7$

f) $(\sqrt{2y} + 7)(\sqrt{2y} - 1) = \sqrt{4y^2} - \sqrt{2y} + 7\sqrt{2y} - 7 = 2y + 6\sqrt{2y} - 7$

g) $(\sqrt{x-2} + 3)(\sqrt{x-2} - 3) = \sqrt{(x-2)^2} - 3\sqrt{x-2} + 3\sqrt{x-2} - 9 = (x-2) - 9 = x - 11$

❖ *Dividing radical expressions - same index*

In this section we divide radical expressions with the same index. To proceed, divide the numbers or expressions outside the radical together and the numbers or expressions inside the radicals together. If the final answer has a radical index two in the denominator of the fraction, then simplify it by removing the radical in the denominator. This method is called **rationalizing the denominator**. It is done by multiplying the numerator and the denominator of the fraction by the same radical in the denominator to obtain a perfect square radical in the denominator.

Quotient Rule for dividing

$$\sqrt[n]{\frac{a}{b}} = \frac{\sqrt[n]{a}}{\sqrt[n]{b}} \ , \sqrt[n]{b} \neq 0$$

If the denominator has a binomial with a radical index two, then multiply the numerator and denominator by the **conjugate** (same number but opposite signs in the middle) to rationalize the denominator.

Examples of Conjugates

$$(a + b) \rightarrow (a - b)$$

$$(\sqrt{a} - \sqrt{b}) \rightarrow (\sqrt{a} + \sqrt{b})$$

$$(a + \sqrt{b}) \rightarrow (a - \sqrt{b})$$

Example 4

Divide the radical expressions.

a) $\sqrt{\dfrac{9}{49}} = \dfrac{\sqrt{9}}{\sqrt{49}} = \dfrac{3}{7}$

b) $\dfrac{\sqrt{18}}{3} = \dfrac{\sqrt{9 \cdot 2}}{3} = \dfrac{3\sqrt{2}}{3} = \sqrt{2}$

c) $\dfrac{\sqrt{54}}{\sqrt{6}} = \dfrac{\sqrt{9 \cdot 6}}{\sqrt{6}} = \dfrac{3\sqrt{6}}{\sqrt{6}} = 3$

d) $\sqrt[3]{\dfrac{27}{64}} = \dfrac{\sqrt[3]{27}}{\sqrt[3]{64}} = \dfrac{3}{4}$

e) $\dfrac{\sqrt{48x^3}}{\sqrt{3}} = \dfrac{\sqrt{16 \cdot 3 \, x^2 \, x}}{\sqrt{3}} = \dfrac{4x\sqrt{3x}}{\sqrt{3}} \cdot \dfrac{4x\sqrt{3}\sqrt{x}}{\sqrt{3}} = 4x\sqrt{x}$

f) $\dfrac{\sqrt{25}}{\sqrt{7}} = \dfrac{5}{\sqrt{7}} = \dfrac{5}{\sqrt{7}} \cdot \dfrac{\sqrt{7}}{\sqrt{7}} = \dfrac{5\sqrt{7}}{\sqrt{49}} = \dfrac{5\sqrt{7}}{7}$

g) $\dfrac{3}{4+\sqrt{5}} = \dfrac{(3)}{(4+\sqrt{5})} \cdot \dfrac{(4-\sqrt{5})}{(4-\sqrt{5})} = \dfrac{12-3\sqrt{5}}{16-5} = \dfrac{12-3\sqrt{5}}{11}$

h) $\dfrac{x}{\sqrt{3}-1} = \dfrac{(x)}{(\sqrt{3}-1)} \cdot \dfrac{(\sqrt{3}+1)}{(\sqrt{3}+1)} = \dfrac{x\sqrt{3}+x}{3-1} = \dfrac{(1+\sqrt{3})\,x}{2}$

i) $\dfrac{4}{\sqrt{5}-\sqrt{2}} = \dfrac{(4)}{(\sqrt{5}-\sqrt{2})} \cdot \dfrac{(\sqrt{5}+\sqrt{2})}{(\sqrt{5}+\sqrt{2})} = \dfrac{4\sqrt{5}+4\sqrt{2}}{5-2} = \dfrac{4\sqrt{5}+4\sqrt{2}}{3}$

Section 5.4 – Multiplying and dividing radical expressions.

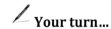 **Your turn...**

Multiply the radical expressions.

1. $\sqrt{2} \cdot \sqrt{6}$

2. $\sqrt{4} \cdot \sqrt{5}$

3. $\sqrt[3]{5} \cdot \sqrt[3]{16}$

4. $\left(2\sqrt{y}\right)\left(5\sqrt{2y}\right)$

5. $\left(\frac{1}{3}\sqrt{3}\right)\left(\frac{3}{5}\sqrt{6}\right)$

6. $\sqrt{4}\left(1 + \sqrt{2}\right)$

7. $\sqrt{2}\left(2\sqrt{8} - \sqrt{20}\right)$

8. $\sqrt{5}\left(\sqrt{8} + \sqrt{6}\right)$

9. $\left(\sqrt{2} + \sqrt{5}\right)\left(\sqrt{4} + \sqrt{3}\right)$

10. $\left(1 + \sqrt{6}\right)\left(\sqrt{3} - 4\right)$

11. $\left(\sqrt{5} + 7\right)\left(\sqrt{5} - 7\right)$

12. $\left(\sqrt{3} + 7\right)\left(\sqrt{3} + 1\right)$

13. $\left(\sqrt{x-1} + 4\right)\left(\sqrt{x-1} - 4\right)$

14. $\left(\sqrt{x+2} + 5\right)\left(\sqrt{x+2} - 5\right)$

Divide the radical expressions.

15. $\sqrt{\dfrac{16}{25}}$

16. $\dfrac{\sqrt{48}}{4}$

17. $\dfrac{\sqrt{72}}{\sqrt{2}}$

18. $\sqrt[3]{\dfrac{8}{125}}$

19. $\dfrac{\sqrt{45x^5}}{\sqrt{5}}$

20. $\dfrac{\sqrt{81}}{\sqrt{11}}$

21. $\dfrac{1}{7-\sqrt{3}}$

For more practice, access the online homework. See your syllabus for details.

> ### Section 5.5 – Solving Radical Equations
>
> ❖ Solving radical equations

❖ *Solving radical equations*

In order to solve radical equations, first isolate the radical expressions if necessary. Then look at the index. If the index is 2, raise both sides of the equation to the second power. If the index is 3, raise both sides of the equation to the third power. If the index is 4, raise both sides of the equation to the fourth power and so on. What happens is that the radical sign goes away because the expressions have the same index and power. Once the radical sign goes away or turns to 1, solve the equations. *Also, it is necessary to check the answer(s) in the original equation.*

Example 1

Solve the radical equation.

$\sqrt{x} = 9$

$\left(\sqrt{x}\right)^2 = (9)^2$ (Raise both sides to the second power.)

$\left(x^{\frac{1}{2}}\right)^2 = 81$ (Rewrite as rational exponent.)

$x = 81$ (Power turns to 1 on the left side.)

Check: $\sqrt{81} = 9$

Example 2

Solve the radical equation.

$\sqrt[3]{x} = -4$

$\left(\sqrt[3]{x}\right)^3 = (-4)^3$ (Raise both sides to the third power.)

$\left(x^{\frac{1}{3}}\right)^3 = -64$ (Rewrite as rational exponent.)

$x = -64$ (Power turns to 1 on the left side.)

Check: $\sqrt[3]{-64} = -4$

Example 3

Solve the radical equation.

$\sqrt[4]{x} = 3$

$\left(\sqrt[4]{x}\right)^4 = (3)^4$ (Raise both sides to the fourth power.)

$\left(x^{\frac{1}{4}}\right)^4 = 81$ (Rewrite as rational exponent.)

$x = 81$ (Power turns to 1 on the left side.)

Check: $\sqrt[4]{81} = 3$

Example 4

Solve the radical equation.

$\sqrt{x} + 4 = 7$

First isolate the radical expression.

$\sqrt{x} = 7 - 4$ (Subtract 4 from both sides.)

$\sqrt{x} = 3$ (Combine like terms.)

$\left(\sqrt{x}\right)^2 = (3)^2$ (Raise both sides to the second power.)

$\left(x^{\frac{1}{2}}\right)^2 = 9$ (Rewrite as rational exponent.)

$x = 9$ (Power turns to 1 on the left side.)

Check: $\sqrt{x} + 4 = 7 \rightarrow \sqrt{9} + 4 = 7 \rightarrow 3 + 4 = 7$

Example 5

Solve the radical equation.

$\sqrt{x - 3} + 6 = 10$

First isolate the radical expression.

$\sqrt{x - 3} = 10 - 6$ (Subtract 6 from both sides.)

$\sqrt{x - 3} = 4$ (Combine like terms.)

$$\left(\sqrt{x-3}\right)^2 = (4)^2 \qquad \text{(Raise both sides to the second power.)}$$

$$\left((x-3)^{\frac{1}{2}}\right)^2 = 16 \qquad \text{(Rewrite as rational exponent.)}$$

$$x - 3 = 16 \qquad \text{(Power turns to 1 on the left side.)}$$

$$x = 16 + 3 \qquad \text{(Add 3 to both sides.)}$$

$$x = 19$$

Check: $\sqrt{x-3} + 6 = 10 \rightarrow \sqrt{19-3} + 6 = 10 \rightarrow \sqrt{16} + 6 = 10 \rightarrow 4 + 6 = 10$

Example 6

Solve the radical equation.

$$\sqrt{x-4} + 5 = 3$$

First isolate the radical expression.

$$\sqrt{x-4} = 3 - 5 \qquad \text{(Subtract 5 from both sides.)}$$

$$\sqrt{x-4} = -2 \qquad \text{(False statement with even index – No solution)}$$

$$\left(\sqrt{x-4}\right)^2 = (-2)^2 \qquad \text{(If solved, the solution will not work. Raise both sides to the second power.)}$$

$$\left((x-4)^{\frac{1}{2}}\right)^2 = 4 \qquad \text{(Rewrite as rational exponent.)}$$

$$x - 4 = 4 \qquad \text{(Power turns to 1 on the left side.)}$$

$$x = 4 + 4 \qquad \text{(Add 4 to both sides.)}$$

$$x = 8$$

Check: $\sqrt{x-4} + 5 = 3 \rightarrow \sqrt{8-4} + 5 = 3 \rightarrow \sqrt{4} + 5 = 3 \rightarrow 2 + 5 = 3$

$7 \neq 3$. **No solution.**

Example 7

Solve the radical equation.

$$\sqrt[3]{x+7} + 2 = -1$$

First isolate the radical expression.

$\sqrt[3]{x+7} = -1 - 2$ (Subtract 2 from both sides.)

$\sqrt[3]{x+7} = -3$ (Combine like terms.)

$\left(\sqrt[3]{x+7}\right)^3 = (-3)^3$ (Raise both sides to the third power.)

$\left((x+7)^{\frac{1}{3}}\right)^3 = -27$ (Rewrite as rational exponent.)

$x + 7 = -27$ (Power turns to 1 on the left side.)

$x = -27 - 7$ (Subtract 7 from both sides.)

$x = -34$

Check: $\sqrt[3]{x+7} + 2 = -1 \rightarrow \sqrt[3]{-34+7} + 2 = -1 \rightarrow \sqrt[3]{-27} + 2 = -1$

$-3 + 2 = -1$

Example 8

Solve the radical equation.

$\sqrt{2x+9} = \sqrt{x+7}$

The radical expressions are already isolated.

$\left(\sqrt{2x+9}\right)^2 = \left(\sqrt{x+7}\right)^2$ (Raise both sides to the second power.)

$\left((2x+9)^{\frac{1}{2}}\right)^2 = \left((x+7)^{\frac{1}{2}}\right)^2$ (Rewrite as rational exponent.)

$2x + 9 = x + 7$ (Power turns to 1 on both sides.)

$2x - x + 9 = 7$ (Subtract x from both sides.)

$x + 9 = 7 \rightarrow x = 7 - 9$ (Combine like terms.)

$x = -2$

Check: $\sqrt{2x+9} = \sqrt{x+7} \rightarrow \sqrt{2(-2)+9} = \sqrt{-2+7} \rightarrow \sqrt{-4+9} = \sqrt{-2+7} \rightarrow \sqrt{5} = \sqrt{5}$

Example 9

Solve the radical equation.

$$\sqrt[3]{x - 7} = \sqrt[3]{4x + 11}$$

The radical expressions are already isolated.

$$\left(\sqrt[3]{x - 7}\right)^3 = \left(\sqrt[3]{4x + 11}\right)^3 \qquad \text{(Raise both sides to the third power.)}$$

$$\left((x - 7)^{\frac{1}{3}}\right)^3 = \left((4x + 11)^{\frac{1}{3}}\right)^3 \qquad \text{(Rewrite as rational exponent.)}$$

$$x - 7 = 4x + 11 \qquad \text{(Power turns to 1 on both sides.)}$$

$$x - 4x - 7 = 11 \qquad \text{(Subtract } 4x \text{ from both sides.)}$$

$$-3x - 7 = 11 \rightarrow -3x = 11 + 7 \qquad \text{(Combine like terms.)}$$

$$-3x = 18 \rightarrow x = \frac{18}{-3} \rightarrow x = -6$$

Check: $\sqrt[3]{x - 7} = \sqrt[3]{4x + 11} \rightarrow \sqrt[3]{-6 - 7} = \sqrt[3]{4(-6) + 11} \rightarrow \sqrt[3]{-13} = \sqrt[3]{-13}$

Example 10

Solve the radical equation.

$$\sqrt[5]{5x - 17} = \sqrt[5]{3x - 7}$$

The radical expressions are already isolated.

$$\left(\sqrt[5]{5x - 17}\right)^5 = \left(\sqrt[5]{3x - 7}\right)^5 \qquad \text{(Raise both sides to the fifth power.)}$$

$$\left((5x - 17)^{\frac{1}{5}}\right)^5 = \left((3x - 7)^{\frac{1}{5}}\right)^5 \qquad \text{(Rewrite as rational exponent.)}$$

$$5x - 17 = 3x - 7 \qquad \text{(Power turns to 1 on both sides.)}$$

$$5x - 3x - 17 = -7 \qquad \text{(Subtract } 3x \text{ from both sides.)}$$

$$2x - 17 = -7 \qquad \text{(Combine like terms.)}$$

$$2x = -7 + 17 \qquad \text{(Add 17 to both sides.)}$$

$$2x = 10 \rightarrow x = 5$$

Check: $\sqrt[5]{5x - 17} = \sqrt[5]{3x - 7} \rightarrow \sqrt[5]{5(5) - 17} = \sqrt[5]{3(5) - 7} \rightarrow \sqrt[5]{25 - 17} = \sqrt[5]{15 - 7} \rightarrow \sqrt[5]{8} = \sqrt[5]{8}$

Example 11

Solve the radical equation.

$$\sqrt{x + 4} + 2 = x$$

First isolate the radical expression.

$\sqrt{x + 4} = x - 2$	(Subtract 2 from both sides.)
$\left(\sqrt{x + 4}\right)^2 = (x - 2)^2$	(Raise both sides to the second power.)
$x + 4 = (x - 2)(x - 2)$	(Expand the right side of the equation.)
$x + 4 = x^2 - 4x + 4$	(Use FOIL on the right side of the equation.)

Set the equation equal to zero in order to solve for x.

$0 = x^2 - 4x + 4 - x - 4$	(Subtract both x and 4 from both sides.)
$0 = x^2 - 5x$	(Combine like terms.)
$0 = x(x - 5)$	(Factor.)
$x = 0 \qquad x - 5 = 0 \rightarrow x = 5$	(Solve each factor.)

Check: $\sqrt{x + 4} + 2 = x \rightarrow \sqrt{0 + 4} + 2 = 0 \rightarrow 4 \neq 0$

Check: $\sqrt{x + 4} + 2 = x \rightarrow \sqrt{5 + 4} + 2 = 5 \rightarrow 3 + 2 = 5 \rightarrow 5 = 5$

There is one solution and it is $x = 5$.

Example 12

Solve the radical equation.

$$\sqrt{x - 2} = 1 + \sqrt{x - 7}$$

We cannot isolate the radical for this example. Let us try to clear the radical sign on the left side of the equation.

$\left(\sqrt{x - 2}\right)^2 = \left(1 + \sqrt{x - 7}\right)^2$	(Raise both sides to the second power.)
$x - 2 = (1 + \sqrt{x - 7})(1 + \sqrt{x - 7})$	(Expand the right side of the equation.)
$x - 2 = 1 + 2\sqrt{x - 7} + x - 7$	(Use FOIL.)
$x - 2 - 1 - x + 7 = 2\sqrt{x - 7}$	(Isolate the radical on the right side of the equation.)

$4 = 2\sqrt{x-7}$ (Combine like terms.)

$2 = \sqrt{x-7}$ (Divide both sides by 2.)

$(2)^2 = \left(\sqrt{x-7}\right)^2$ (Raise both sides to the second power.)

$4 = x - 7$ (Radical sign is cleared.)

$4 + 7 = x$ (Add 7 to both sides.)

$x = 11$

Check: $\sqrt{x-2} = 1 + \sqrt{x-7} \rightarrow \sqrt{11-2} = 1 + \sqrt{11-7} \rightarrow \sqrt{9} = 1 + \sqrt{4} \rightarrow 3 = 3$

Example 13

If nine is added to the square root of a number and 5, the result is 12. Find the number.

$\sqrt{x+5} + 9 = 12$

Isolate the radical expression.

$\sqrt{x+5} = 12 - 9$ (Subtract 9 from both sides.)

$\sqrt{x+5} = 3$ (Combine like terms.)

$\left(\sqrt{x+5}\right)^2 = (3)^2$ (Raise both sides to the second power.)

$\left((x+5)^{\frac{1}{2}}\right)^2 = 9$ (Rewrite as rational exponent.)

$x + 5 = 9$ (Power turns to 1 on the left side.)

$x = 9 - 5 \rightarrow \quad x = 4$

Check: $\sqrt{x+5} + 9 = 12 \rightarrow \sqrt{4+5} + 9 = 12 \rightarrow \sqrt{9} + 9 = 12 \rightarrow 3 + 9 = 12$

Example 14

The formula for the radius of a cylinder is given by $\sqrt{\dfrac{V}{\pi h}} = r$. The height of a certain cylinder is 10 cm and the radius is 5 cm. Use the given formula to find the volume of the cylinder. (Use 3.1416 as an approximation for π).

To find the volume, we use the given formula:

$$\sqrt{\frac{V}{\pi h}} = r$$

$$\sqrt{\frac{V}{3.1416(10)}} = 5 \qquad \text{(Replace } \pi \text{ with 3.1416, the height with 10 and the radius with 5.)}$$

To get the volume, we need to clear the radical sign.

$$\left(\sqrt{\frac{V}{31.416}} \right)^2 = (5)^2 \qquad \text{(Raise both sides to the second power.)}$$

$$\frac{V}{31.416} = 25 \qquad \text{(Radical sign is cleared.)}$$

$$V = 25 \cdot 31.416 \rightarrow V = 785.4 \; cm^3 \qquad \text{(Multiply both sides by 31.416 to isolate V.)}$$

Example 15

The formula for the radius of a sphere is given by $\sqrt[3]{\frac{3V}{4\pi}} = r$. The radius of a certain sphere is 6 cm. Use the given formula to find the volume of the sphere. (Use 3.1416 as an approximation for π).

To find the volume, we use the given formula:

$$\sqrt[3]{\frac{3V}{4\pi}} = r$$

$$\sqrt[3]{\frac{3V}{4(3.1416)}} = 6 \qquad \text{(Replace } \pi \text{ with 3.1416 and the radius with 6.)}$$

To get the volume, we need to clear the radical sign.

$$\left(\sqrt[3]{\frac{3V}{4(3.1416)}} \right)^3 = (6)^3 \qquad \text{(Raise both sides to the third power.)}$$

$$\frac{3V}{12.5664} = 216 \qquad \text{(Radical sign is cleared.)}$$

$$3V = 216 \cdot 12.5664 \qquad \text{(Multiply both sides by 12.5664 to clear the fraction.)}$$

$$V = \frac{(216)(12.5664)}{3} \qquad \text{(Divide both sides by 3 to isolate V.)}$$

$$V = 904.8 \; cm^3$$

Section 5.5 – Solving radical equations

 **Your turn...**

Solve.

1. $\sqrt{x} = 6$

2. $\sqrt[3]{x} = -5$

3. $\sqrt[4]{x} = 4$

4. $\sqrt{x} + 3 = 6$

5. $\sqrt{x-2} + 5 = 9$

6. $\sqrt{x-3} + 4 = 2$

7. $\sqrt[3]{x+9} + 3 = -2$

8. $\sqrt{4x+7} = \sqrt{x+5}$

9. $\sqrt[3]{x-3} = \sqrt[3]{2x+10}$

10. $\sqrt[4]{2x-17} = \sqrt[4]{4x-1}$

11. $\sqrt{x+9} + 3 = x$

12. $\sqrt{x-3} = 1 + \sqrt{x-10}$

13. If ten is added to the square root of a number and 3, the result is 14. Find the number.

14. The formula for the radius of a cylinder is given by $\sqrt{\dfrac{V}{\pi h}} = $ r. The height of a certain cylinder is 11 cm and the radius is 4 cm. Use the given formula to find the volume of the cylinder. (Use 3.1416 as an approximation for π).

15. The formula for the radius of a sphere is given by $\sqrt[3]{\dfrac{3V}{4\pi}} = r$. The radius of a certain sphere is 3 cm. Use the given formula to find the volume of the sphere. (Use 3.1416 as an approximation for π).

For more practice, access the online homework. See your syllabus for details.

Section 5.6 – Working with Complex Numbers

- ❖ Expressions with "i" or imaginary numbers
- ❖ Higher power of "i"
- ❖ Complex numbers
  - o Identifying numbers as real or imaginary
  - o Adding and subtracting complex numbers
  - o Multiplying complex numbers
  - o Dividing complex numbers

❖ *Expressions with "i" or imaginary numbers*

Simplifying an expression like $\sqrt{-9}$ will not give a real number as seen at the beginning of the chapter. In this section, we can actually simplify similar expressions by expressing it using the letter "i". Let us keep the following in mind as we simplify these expressions.

a) $i^2 = -1$

b) $i = \sqrt{-1}$

Example 1

Simplify.

a) $\sqrt{-36} = \sqrt{-1}\,\sqrt{36} = i(6) = 6i$

b) $\sqrt{-20} = \sqrt{-1}\,\sqrt{20} = i\sqrt{4 \cdot 5} = i(2)\sqrt{5} = 2i\sqrt{5}$

c) $\sqrt{-7} = \sqrt{-1}\,\sqrt{7} = i\,\sqrt{7}$

Example 2

Multiply.

Note: One cannot multiply radicands when they are negative.

a) $\sqrt{-2}\,\sqrt{-5} = \left(\sqrt{-1}\,\sqrt{2}\,\right)\left(\sqrt{-1}\,\sqrt{5}\,\right) = \left(i\sqrt{2}\right)\left(i\sqrt{5}\right) = i^2\sqrt{10} = -1\sqrt{10} = -\sqrt{10}$

b) $\sqrt{-3}\,\sqrt{-6} = \left(\sqrt{-1}\,\sqrt{3}\right)\left(\sqrt{-1}\,\sqrt{6}\right) = \left(i\sqrt{3}\right)\left(i\sqrt{6}\right) = i^2\sqrt{18} = -1\sqrt{9\cdot 2} = -3\sqrt{2}$

c) $\sqrt{-7}\,\sqrt{-7} = \left(\sqrt{-1}\,\sqrt{7}\,\right)\left(\sqrt{-1}\,\sqrt{7}\,\right) = \left(i\sqrt{7}\right)\left(i\sqrt{7}\right) = i^2\sqrt{49} = -1(7) = -7$

Example 3

Divide.

a) $\dfrac{\sqrt{-4}}{\sqrt{-16}} = \dfrac{\sqrt{-1}\,\sqrt{4}}{\sqrt{-1}\,\sqrt{16}} = \dfrac{2i}{4i} = \dfrac{2}{4} = \dfrac{1}{2}$

b) $\dfrac{\sqrt{-45}}{\sqrt{-80}} = \dfrac{\sqrt{-1}\,\sqrt{9\cdot 5}}{\sqrt{-1}\,\sqrt{16\cdot 5}} = \dfrac{3i\sqrt{5}}{4i\sqrt{5}} = \dfrac{3}{4}$

c) $\dfrac{\sqrt{-3}}{\sqrt{-7}} = \dfrac{\sqrt{-1}\,\sqrt{3}}{\sqrt{-1}\,\sqrt{7}} = \dfrac{i\sqrt{3}}{i\sqrt{7}} = \dfrac{\sqrt{3}}{\sqrt{7}}$ or $\dfrac{\sqrt{21}}{7}$ ($\dfrac{\sqrt{3}}{\sqrt{7}}$ is multiplied by $\dfrac{\sqrt{7}}{\sqrt{7}}$ or the denominator is rationalized)

❖ *Higher powers of i*

To find patterns that will help us, when finding higher powers of i, we can use these relationships $i^2 = -1$ and $i = \sqrt{-1}$.

For example, to find i^3, we can rewrite it this way.

$$i^3 = i^2 \cdot i = (-1) \cdot i = -i$$

Using similar techniques, we can come up with the following patterns as shown in the table below.

$i^0 = 1$	$i^1 = i$	$i^2 = -1$	$i^3 = -i$
$i^4 = 1$	$i^5 = i$	$i^6 = -1$	$i^7 = -i$
$i^8 = 1$	$i^9 = i$	$i^{10} = -1$	$i^{11} = -i$
$i^{12} = 1$	$i^{13} = i$	$i^{14} = -1$	$i^{15} = -i$

Example 4

Find higher powers of i.

a) $i^{11} = (i^{10})(i) = (-1)(i) = -i$ OR $i^{11} = (i^2)^5(i) = (-1)(i) = -i$

b) $i^{36} = (i^6)^6 = (-1)^6 = 1$ OR $i^{36} = (i^2)^{18} = (-1)^{18} = 1$

c) $i^{72} = (i^{70})(i^2) = (i^{10})^7(i^2) = (-1)^7(-1) = 1$ OR $i^{72} = (i^2)^{36} = (-1)^{36} = 1$

d) $i^{-16} = \frac{1}{i^{16}} = \frac{1}{(i^4)^4} = \frac{1}{(1)^4} = 1$ OR $i^{-16} = \frac{1}{i^{16}} = \frac{1}{(i^2)^8} = \frac{1}{(-1)^4} = \frac{1}{1} = 1$

❖ *Complex numbers*

A complex number is a number, which is in the **standard form** of $a + bi$, where "a" and "b" are real numbers and i is the imaginary unit.

o Identifying numbers as real or imaginary

A complex number is a real number if b is 0. It is a pure imaginary number is $a = 0$ and b ≠ 0. Below is a table that shows the different types of complex numbers.

Types	Form
Imaginary Unit	i
Complex Number	$a + bi$; a and b are real numbers
• Imaginary Number	$a + bi$; $b \neq 0$
• Pure Imaginary Number	$0 + bi = bi,\ b \neq 0$
Real number	$a + 0i = a$

o Adding and subtracting complex numbers

To add or subtract complex numbers, simply add or subtract their real parts and then add or subtract their imaginary parts.

Example 5

Add or subtract the complex numbers. Write the final answer in standard form.

a) $(3 + 4i) + (-1 + 5i) = 3 + 4i - 1 + 5i = 2 + 9i$

b) $7i - (3 + i) = 7i - 3 - i = -3 + 6i$

c) $(5 + 9i) - (-4) = 5 + 9i + 4 = 9 + 9i$

d) $(6 + 2i) - (12 - i) = 6 + 2i - 12 + i = -6 + 3i$

o <u>Multiplying complex numbers</u>

To multiply complex numbers, we use the same rules that we normally use when multiplying non-complex numbers. We use the power rule, distributive property, FOIL method. For complex numbers, we use the relationship $i^2 = -1$ to simplify.

Example 6

Multiply the complex numbers. Write the final answer in standard form.

a) $(5i)(2i) = 10\,i^2 = 10\,(-1) = -10 \quad \text{or} -10 + 0i$

b) $(-3i)(4i) = -12\,i^2 = 12\,(-1) = 12 \quad \text{or } 12 + 0i$

c) $(7i)(1 + 5i) = 7i + 35i^2 = 7i + 35(-1) = 7i - 35 = -35 + 7i$

d) $(2 + 9i)(2 - 9i) = 4 - 18i + 18i - 81i^2 = 4 - 81(-1) = 85 \ \text{ or } 85 + 0i$

e) $(4 - 5i)(1 + i) = 4 + 4i - 5i - 5i^2 = 4 - i - 5(-1) = 9 - i$

f) $(11 - i)^2 = (11 - i)(11 - i) = 121 - 11i - 11i + i^2 = 121 - 22i - 1 = 120 - 22i$

o <u>Dividing complex numbers</u>

To divide complex numbers, multiply both the numerator and denominator by the conjugate of the complex number in the denominator.

Complex Conjugate

$$(a + bi) \rightarrow (a - bi)$$

The product or multiplication of two complex conjugates is a real number.

$$(a + bi)(a - bi) = a^2 + b^2$$

Example 7

Divide the complex numbers. Write the final answer in standard form,

a) $\dfrac{2-i}{4+i} = \dfrac{(2-i)}{(4+i)} \cdot \dfrac{(4-i)}{(4-i)} = \dfrac{8-2i-4i+i^2}{16-i^2} = \dfrac{8-6i-1}{16-(-1)} = \dfrac{7-6i}{17}$ or $\dfrac{7}{17} - \dfrac{6}{17}i$

b) $\dfrac{1+2i}{5-3i} = \dfrac{(1+2i)}{(5-3i)} \cdot \dfrac{(5+3i)}{(5+3i)} = \dfrac{5+3i+10i+6i^2}{25-9i^2} = \dfrac{5+13i-6}{25-9(-1)} = \dfrac{-1+13i}{34}$ or $-\dfrac{1}{34} + \dfrac{13}{34}i$

c) $\dfrac{10}{9i} = \dfrac{(10)}{(9i)} \cdot \dfrac{(-9i)}{(-9i)} = \dfrac{-90i}{-81i^2} = \dfrac{-90i}{-81(-1)} = \dfrac{-90i}{81} = \dfrac{-10i}{9}$ or $0 - \dfrac{10}{9}i$

Note for part c): $9i$ can be written as $0 + 9i$ so the conjugate is $0 - 9i$ or $-9i$.

Section 5.6 - Working with complex numbers

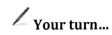 **Your turn...**

Simplify.

1. $\sqrt{-49}$

2. $\sqrt{-40}$

3. $\sqrt{-11}$

Multiply.

4. $\sqrt{-3}\,\sqrt{-7}$

5. $\sqrt{-4}\,\sqrt{-6}$

6. $\sqrt{-3}\,\sqrt{-3}$

Divide.

7. $\dfrac{\sqrt{-16}}{\sqrt{-25}}$

8. $\dfrac{\sqrt{-40}}{\sqrt{-80}}$

9. $\dfrac{\sqrt{-5}}{\sqrt{-6}}$

10. Find higher powers of i.

   a) i^{13}

   b) i^{56}

   c) i^{66}

   d) i^{-24}

11. Perform the indicated operation and write the final answer in standard form.

a) $(6 + 2i) + (5 - 11i)$

b) $2i - (-11 - i)$

c) $(1 + 8i) - (2)$

d) $(2 + 6i) - (4 - i)$

e) $(3i)(10i)$

f) $(-5i)(7i)$

g) $(2i)(1 - 3i)$

h) $(1 + 10i)(1 - 10i)$

i) $(1 - 2i)(3 + 4i)$

j) $(12 + i)^2$

k) $\dfrac{6-i}{2+i}$

l) $\dfrac{2+3i}{1-7i}$

m) $\dfrac{11}{5i}$

For more practice, access the online homework. See your syllabus for details.

Chapter 5: Review of Terms, Concepts, and Formulas ⭐

- **Simplifying radical expressions (Same index/power) -** If the index is even and the expression under the radical is positive, the answer is the positive expression without the radical sign. If the index is even and the expression under the radical is negative, the answer is still a positive expression without the radical sign, $\sqrt{5^2} = 5$; $\sqrt{(-5)^2} = 5$

  If the index is odd and the expression under the radical is positive, the answer is the positive expression without the radical sign. If the index is odd and the expression under the radical is negative, the answer is the negative expression without the radical sign. $\sqrt[3]{4^3} = 4$; $\sqrt[3]{(-4)^3} = -4$

- **Root approximation -** When simplifying numbers that are not perfect squares, cubes, fourths, and fifths, we can use a calculator to find decimal approximations of them. We express the answer using the approximation symbol ($\approx$).

- When simplifying a radical expression, look at the index. If the index is 2, the number must be broken up into 2 factors where one of the factors is a perfect square. If the index is 3, the number must be broken up into 2 factors where one of the factors is a perfect cube. If the index is 4, the number must be broken up into 2 factors where one of the factors is a perfect fourth and so on. Then the perfect square or cube or fourth will be without the radical sign and the other factor will still have the radical sign. [$\sqrt{20} = \sqrt{4 \cdot 5} = \sqrt{4} \cdot \sqrt{5} = 2\sqrt{5}$]

- **Pythagorean Theorem:** $c^2 = a^2 + b^2$ ⭐

- A number is in **scientific notation** if it is expressed in the format $n \times 10^p$. The number n has to be between 1 and 10 (1 digit before the decimal point). The power or exponent is p. The sign of p is positive if the decimal point is moved to the left, and it is negative if moved to the right. The value of p is the number of times the decimal point is moved.

- To convert a number from scientific notation to **standard**, move the decimal point to the right if p is positive. Move the decimal point to the left if p is negative.

- In order to **add/subtract numbers in scientific notation**, the exponents must be the same. If the exponents are the same, keep the exponents and add/subtract the numbers. If the exponents are not the same, make them the same, then keep the exponents and add/subtract the numbers. To make the exponents the same, move the decimal point to the left to increase the exponent or to the right to decrease the exponent.

- In order to **multiply numbers in scientific notation**, the exponents do not need to be the same. Simply multiply the numbers, and **add** the exponents. Make sure the final answer is in scientific notation. (Number between 1 and 10)

- In order to **divide numbers in scientific notation**, the exponents do not need to be the same. Simply divide the numbers, and **subtract** the exponents. Make sure the final answer is in scientific notation. (Number between 1 and 10)

- An expression with a **rational exponent** has a base and an exponent in the form of a fraction or ratio. An example of an expression with a rational exponent is $b^{\frac{n}{m}}$ or $4^{\frac{1}{2}}$. To simplify expressions with rational exponents, convert the expression to a radical expression. Looking at the fraction exponent, the denominator of the fraction becomes the index and the numerator of the fraction becomes the power. Then evaluate the radical expression. An example is $4^{\frac{1}{2}} = \sqrt{4} = 2$.

- If the **rational exponent** is **negative**, before converting to a radical expression, make it positive. An example is $4^{-\frac{1}{2}}$. Take the reciprocal of the expression or write it as 1 over the expression $\left(\frac{1}{4^{\frac{1}{2}}}\right)$. The power becomes positive. Convert to radical as mentioned earlier and evaluate the radical expression. An example is $4^{-\frac{1}{2}} = \frac{1}{4^{\frac{1}{2}}} = \frac{1}{\sqrt{4}} = \frac{1}{2}$.

- When performing **arithmetic operations with rational exponents**, we can use the same rules that are used for integer exponents. Below is a review of the rules and we will let m and n be rational numbers.

Rules of exponents	
Negative exponent Rule	$x^{-n} = \dfrac{1}{x^n}$
Product Rule	$x^m \cdot x^n = x^{m+n}$
Quotient Rule	$\dfrac{x^m}{x^n} = x^{m-n}, \qquad x \neq 0$
Power Rule	$(x^m)^n = x^{m \cdot n}; \; (x)^0 \text{or} \left(\dfrac{x}{y}\right)^0 = 1; \; x, y \neq 0$
Power Rule (Products)	$(xy)^n = x^n \cdot y^n$
Power Rule (Quotients)	$\left(\dfrac{x}{y}\right)^n = \dfrac{x^n}{y^n}; \; \left(\dfrac{x}{y}\right)^{-n} = \left(\dfrac{y}{x}\right)^n, \; y \neq 0$

- To **add and subtract radical expressions**, they must be like radicals. In other words, the expression or number under the radical sign, the radicand, must be the same. They must also have the same index. In this case, just add the coefficients. If the expressions are not like radicals, simplify first, and then add or subtract if possible.

- To **multiply radical expressions** with the same index, multiply the numbers outside the radical together and multiply the number under the radical sign or radicand together. Then if the result needs to be simplified, do so. In the multiplication process, we use the distributive property and the FOIL method when applicable.

- To **divide radical** expressions with the same index, divide the numbers or expressions outside the radical together and the numbers or expressions inside the radicals together. If the final answer has a radical index two in the denominator of the fraction, then simplify it by removing the radical in the denominator. This method is called **rationalizing the denominator**. It is done by multiplying the numerator and the denominator of the fraction by the same radical in the denominator to obtain a perfect square radical in the denominator.

Quotient Rule for dividing

$$\sqrt[n]{\frac{a}{b}} = \frac{\sqrt[n]{a}}{\sqrt[n]{b}} \ , \ \sqrt[n]{b} \neq 0$$

If the denominator has a binomial with a radical index two, then multiply the numerator and denominator by the **conjugate** (same number but opposite signs in the middle) to rationalize the denominator.

Examples of Conjugates

$$(a + b) \rightarrow (a - b)$$

$$(\sqrt{a} - \sqrt{b}) \rightarrow (\sqrt{a} + \sqrt{b})$$

$$(a + \sqrt{b}) \rightarrow (a - \sqrt{b})$$

- To **solve radical equations**, first isolate the radical expressions if necessary. Then look at the index. If the index is 2, raise both sides of the equation to the second power. If the index is 3, raise both sides of the equation to the third power. If the index is 4, raise both sides of the equation to the fourth power and so on. What happens is that the radical sign goes away because the expressions have the same index and power. Once the radical sign goes away or turns to 1, solve the equations. *Also, it is necessary to check the answer(s) in the original equation.*

- **Imaginary numbers**

  o $\sqrt{-1} = i$

  o $-1 = i^2$

  o Higher powers of i

$i^0 = 1$	$i^1 = i$	$i^2 = -1$	$i^3 = -i$
$i^4 = 1$	$i^5 = i$	$i^6 = -1$	$i^7 = -i$
$i^8 = 1$	$i^9 = i$	$i^{10} = -1$	$i^{11} = -i$
$i^{12} = 1$	$i^{13} = i$	$i^{14} = -1$	$i^{15} = -i$

- A **complex number** is a number, which is in the **standard form** of $a + bi$, where "a" and "b" are real numbers and i is the imaginary unit. A complex number is a real number if b is 0. It is a pure imaginary number is $a = 0$ and b ≠ 0.

Types	Form
Imaginary Unit	i
Complex Number	$a + bi$; a and b are real numbers
• Imaginary Number	$a + bi$; $b \neq 0$
• Pure Imaginary Number	$0 + bi = bi$, $b \neq 0$
Real number	$a + 0i = a$

- To **add or subtract complex numbers**, simply add or subtract their real parts and then add or subtract their imaginary parts.

- To **multiply complex numbers**, we use the same rules that we normally use when multiplying non- complex numbers. We use the power rule, distributive property, FOIL method. For complex numbers, we use the relationship $i^2 = -1$ to simplify.

- To **divide complex numbers**, multiply both the numerator and denominator by the conjugate of the complex number in the denominator.

Complex Conjugate

$$(a + bi) \rightarrow (a - bi)$$

The product or multiplication of two complex conjugates is a real number.

$$(a + bi)(a - bi) = a^2 + b^2$$

CHAPTER 6

QUADRATIC EQUATIONS & FUNCTIONS

Section 6.1 - Solving Quadratic Equations - Factoring

❖ Solving quadratic equations by factoring

❖ *Solving quadratic equations by factoring*

When solving an equation by factoring, first make sure it is **equal to zero**, and then factor the equation using the techniques learned in section 3.5. Then set each factor equal to zero (use the zero-product rule) and solve. *The solutions or the x-values found are actually x-intercepts.*

Example 1

Solve by factoring.

$x^2 + 7x + 10 = 0$

$(x + 5)(x + 2) = 0$ (Factor.)

$x + 5 = 0 \rightarrow \boldsymbol{x = -5}$ (Set each factor equal to 0 then solve.)

$x + 2 = 0 \rightarrow \boldsymbol{x = -2}$ (Set each factor equal to 0 then solve.)

Example 2

Solve by factoring.

$x^2 - 2x - 63 = 0$

$(x - 9)(x + 7) = 0$ (Factor.)

$x - 9 = 0 \rightarrow \boldsymbol{x = 9}$ (Set each factor equal to 0 then solve.)

$x + 7 = 0 \rightarrow \boldsymbol{x = -7}$ (Set each factor equal to 0 then solve.)

Example 3

Solve by factoring.

$x^2 - 8x + 12 = 0$

$(x - 6)(x - 2) = 0$ (Factor.)

$x - 6 = 0 \rightarrow \boldsymbol{x = 6}$ (Set each factor equal to 0 then solve.)

$x - 2 = 0 \rightarrow \boldsymbol{x = 2}$ (Set each factor equal to 0 then solve.)

Example 4

Solve by factoring.

$2x^2 + 7x + 6 = 0$

$2x^2 + 7x + 6$　　　　　　　$(2 \cdot 6 = 12;\ 4 \cdot 3 = 12;\ 4 + 3 = 7)$

$2x^2 + \mathbf{4x} + \mathbf{3x} + 6$　　　　(Rewrite the middle term using 4 and 3 – order does not matter.)

$2x^2 + 4x \quad + 3x + 6$　　　(Factor by grouping.)

$2x(x + 2) \quad + 3(x + 2)$　　　(The binomial $x + 2$ is common, factor it out.)

$(x + 2)(2x + 3) = 0$　　　(Final factoring)

$x + 2 = 0 \rightarrow x = -2$　　　(Solve each factor.)

$2x + 3 = 0 \rightarrow x = -\dfrac{3}{2}$　　(Solve each factor.)

Example 5

Solve by factoring.

$6x^2 - 13x + 5 = 0$

$6x^2 - 13x + 5$　　　　　　$(6 \cdot 5 = 30;\ -3 \cdot -10 = 30;\ -3 + (-10) = -13)$

$6x^2 - \mathbf{3x} - \mathbf{10x} + 5$　　　(Rewrite the middle term using -3 and -10 – order does not matter.)

$6x^2 - 3x \quad - 10x + 5$　　(Factor by grouping.)

$3x(2x - 1) \quad - 5(2x - 1)$　　(The binomial $2x - 1$ is common, factor it out.)

$(2x - 1)(3x - 5) = 0$　　　(Final factoring)

$2x - 1 = 0 \rightarrow x = \dfrac{1}{2}$　　　(Solve each factor.)

$3x - 5 = 0 \rightarrow x = \dfrac{5}{3}$　　　(Solve each factor.)

Example 6

Solve by factoring.

$4x^2 - 4x - 15 = 0$

$4x^2 - 4x - 15$	$(4 \cdot -15 = -60; \ -10 \cdot 6 = -60; \ -10 + 6 = -4)$
$4x^2 - 10x + 6x - 15$	(Rewrite the middle term using -10 and 6– order does not matter.)
$4x^2 - 10x \quad + 6x - 15$	(Factor by grouping.)
$2x(2x - 5) \ + 3(2x - 5)$	(The binomial $2x - 5$ is common, factor it out.)
$(2x - 5)(2x + 3) = 0$	(Final factoring)
$2x - 5 = 0 \rightarrow x = \dfrac{5}{2}$	(Solve each factor.)
$2x + 3 = 0 \rightarrow x = -\dfrac{3}{2}$	(Solve each factor.)

Example 7

Solve by factoring.

$27y^2 + 54y + 27 = 0$

$27(y^2 + 2y + 1)$	(Factor out first a GCF.)
$27(y^2 + 2y + 1)$	(1 and 1; $1 \cdot 1 = 2$; $1 + 1 = 2$)
$27(y + 1)(y + 1) = 0$	(Final factoring)
$27 \neq 0$	(Solve each factor.)
$y + 1 = 0 \rightarrow y = -1$	(Solve the repeated factor.)

Example 8

Solve by factoring.

$81x^2 - 36 = 0$

$(9x + 6)(9x - 6) = 0$	(Factor.)
$9x + 6 = 0 \rightarrow x = -\dfrac{6}{9} = -\dfrac{2}{3}$	(Solve each factor.)

$$9x - 6 = 0 \rightarrow x = \frac{6}{9} \rightarrow \boldsymbol{x = \frac{2}{3}} \qquad \text{(Solve each factor.)}$$

Example 9

Solve by factoring.

$$\frac{4}{49}x^2 - \frac{9}{25} = 0$$

$$\left(\frac{2}{7}x + \frac{3}{5}\right)\left(\frac{2}{7}x - \frac{3}{5}\right) = 0 \qquad \text{(Factor.)}$$

$$\frac{2}{7}x + \frac{3}{5} = 0 \qquad \text{(Solve the first factor.)}$$

The LCD is 35. Multiply all the numerators by 35 to clear the fractions.

$$\frac{2(35)}{7}x + \frac{3(35)}{5} = 0(\boldsymbol{35})$$

$$10x + 21 = 0 \rightarrow \boldsymbol{x = -\frac{21}{10}}$$

$$\frac{2}{7}x - \frac{3}{5} = 0 \qquad \text{(Solve the second factor.)}$$

$$\frac{2(35)}{7}x - \frac{3(35)}{5} = 0(\boldsymbol{35})$$

$$10x - 21 = 0 \rightarrow \boldsymbol{x = \frac{21}{10}}$$

Note: Below is <u>another way</u> to solve for x using, for example, the first factor.

$$\frac{2}{7}x + \frac{3}{5} = 0$$

$$\frac{2}{7}x = -\frac{3}{5}$$

$$\left(\frac{7}{2}\right)\frac{2}{7}x = -\frac{3}{5}\left(\frac{7}{2}\right)$$

$$\boldsymbol{x = -\frac{21}{10}}$$

Example 10

Solve by factoring.

$$x^2 - 11x = 0$$

$$x(x - 11) = 0 \qquad \text{(Factor.)}$$

$x = 0$ (Solve each factor.)

$x - 11 = 0 \rightarrow x = 11$ (Solve each factor.)

Example 11

Solve by factoring.

The area of a rectangular poster in the atrium of building 1 of the Osceola campus is 36 in^2. The length of the poster is 5 inches more than the width. What are the length and the width of the poster?

$A = 36$

$L = w + 5$

$A = L \cdot w$ (Area of a rectangle)

Replace A with 36 and L with $w + 5$ in the area formula.

$36 = (w + 5) \cdot w \rightarrow 36 = w^2 + 5w$

$w^2 + 5w - 36 = 0$ (Subtract 36 from both sides.)

Factor and set each factor equal to zero.

$(w + 9)(w - 4) = 0$

$w + 9 = 0 \rightarrow w = -9$ in   (Discard – negative measurement)

$w - 4 = 0 \rightarrow w = \mathbf{4}$ **in**

Solve for the length.

$L = w + 5$

$L = 4 + 5 \rightarrow l = \mathbf{9}$ **in**

Example 12

Solve by factoring.

Sophia found out that she will not be able to sell her used intermediate algebra book because the campus is going to use a new edition the following semester. Frustrated, she threw the book up in the air. The function that models this scenario is given by $h(t) = -16t^2 + 32t + 48$, where $h(t)$ is the height of the book above the ground and t is the time in seconds. When will the book hit the ground?

When the book hits the ground, the height is zero. Replace $h(t)$ with 0.

$h(t) = -16t^2 + 32t + 48$

$0 = -16t^2 + 32t + 48$

$\dfrac{0}{-16} = \dfrac{-16}{-16}t^2 + \dfrac{32}{-16}t + \dfrac{48}{-16}$ (Simplify.)

$0 = t^2 - 2t - 3$

$0 = (t - 3)(t + 1)$ (Factor and solve.)

$t - 3 = 0 \rightarrow \boldsymbol{t = 3}$ **seconds**

$t + 1 = 0 \rightarrow t = -1$ (Discard – negative time)

The book will hit the ground after **3 seconds**.

Example 13

Solve by factoring.

Oliver wants to sell college textbooks. The revenue function is given by $R(p) = -10p^2 + 2000p$, where $R(p)$ is the revenue and p is the price in dollars. At what price will the revenue be zero or at what price will Oliver make no money?

$R(p) = -10p^2 + 2000p$

$0 = -10p^2 + 2000p$ (Set the revenue function equal to zero.)

$0 = 10p(-p + 200)$ (Factor out 10p and solve.)

$10p = 0 \rightarrow \boldsymbol{p = 0}$

$-p + 200 = 0 \rightarrow \boldsymbol{p = 200}$

Oliver will make no money if he gives away the textbooks for free or $p = \$0.00$ or if he charges $\$200.00$. (Too much)

Section 6.1 - Solving quadratic equations - factoring

Your turn...

Solve by factoring.

1. $x^2 + 6x + 8 = 0$

2. $x^2 - x - 56 = 0$

3. $x^2 - 9x + 14 = 0$

4. $2x^2 + 7x + 3 = 0$

5. $6x^2 - 17x + 7 = 0$

6. $4x^2 - x - 3 = 0$

7. $15y^2 + 30y + 15 = 0$

8. $100x^2 - 16 = 0$

9. $\frac{4}{25}x^2 - \frac{16}{121} = 0$

10. $x^2 - 13x = 0$

11. The area of a rectangular poster in the atrium of building 1 is 60 in^2. The length of the poster is 7 inches more than the width. What are the length and the width of the poster?

12. Marilyn found out that she will not be able to sell her used college algebra book because the campus is going to use a new edition the following semester. Frustrated, she threw the book up in the air. The function that models this scenario is given by $h(t) = -16t^2 + 48t + 64$, where $h(t)$ is the height of the book above the ground and t is the time in seconds. When will the book hit the ground?

13. Oliver wants to sell college textbooks. The revenue function is given by $R(p) = -15p^2 + 4500p$, where $R(p)$ is the revenue and p is the price in dollars. At what price will the revenue be zero or at what price will Oliver make no money?

For more practice, access the online homework. See your syllabus for details.

❖ *Solving quadratic equations using the square root method*

When solving an equation using the square root method, take the square root of both sides. Add ± in front of the radical sign on the right side. The power is two, so there should be two possible answers. Because the left side has the same index and power, the radical will go away or turns to 1. Then solve the equation by isolating the variable. *The solutions or the x-values found are actually x-intercepts.*

Example 1

Solve using the square root method.

$x^2 = 64$

$\sqrt{x^2} = \pm\sqrt{64}$ (To cancel the square, take the square root of both sides.)

$x = \pm 8$

Example 2

Solve using the square root method.

$x^2 = 24$

$\sqrt{x^2} = \pm\sqrt{24}$ (To cancel the square, take the square root of both sides.)

$x = \pm\sqrt{4 \cdot 6}$ (Simplify.)

$x = \pm 2\sqrt{6}$

Example 3

Solve using the square root method.

$x^2 = 3$

$\sqrt{x^2} = \pm\sqrt{3}$ (To cancel the square, take the square root of both sides.)

$x = \pm\sqrt{3}$

Example 4

Solve using the square root method.

$(x + 1)^2 = 25$

$\sqrt{(x + 1)^2} = \pm\sqrt{25}$ (To cancel the square, take the square root of both sides.)

$x + 1 = \pm 5$

Split into 2 equations and solve.

$x + 1 = 5 \rightarrow x = 5 - 1 \rightarrow x = 4$

$x + 1 = -5 \rightarrow x = -5 - 1 \rightarrow x = -6$

Example 5

Solve using the square root method.

$(x - 5)^2 = \dfrac{4}{9}$

$\sqrt{(x - 5)^2} = \pm\sqrt{\dfrac{4}{9}}$ (To cancel the square, take the square root of both sides.)

$x - 5 = \pm\dfrac{2}{3}$

Split into 2 equations and solve.

$x - 5 = \dfrac{2}{3} \rightarrow x = \dfrac{2}{3} + 5 \rightarrow x = \dfrac{17}{3}$

$x - 5 = -\dfrac{2}{3} \rightarrow x = -\dfrac{2}{3} + 5 \rightarrow x = \dfrac{13}{3}$

Example 6

Solve using the square root method.

$(x - 4)^2 = 7$

$\sqrt{(x - 4)^2} = \pm\sqrt{7}$ (To cancel the square, take the square root of both sides.)

$x - 4 = \pm\sqrt{7}$ (Squared expression is cleared.)

$x = 4 \pm \sqrt{7}$ (Add 4 to both sides.)

Example 7

Solve using the square root method.

$(4x + 7)^2 = 3$

$\sqrt{(4x + 7)^2} = \pm\sqrt{3}$ (To cancel the square, take the square root of both sides.)

$4x + 7 = \pm\sqrt{3}$ (Squared expression is cleared.)

$4x = -7 \pm \sqrt{3}$ (Subtract 7 from both sides.)

$x = \dfrac{-7 \pm \sqrt{3}}{4}$ (Divide both sides by 4.)

Example 8

Solve using the square root method.

$x^2 = -81$

$\sqrt{x^2} = \pm\sqrt{-81}$ (To cancel the square, take the square root of both sides.)

$x = \pm \sqrt{-1} \cdot \sqrt{81}$ (Simplify.)

$x = \pm\, i \cdot 9 \rightarrow x = \pm 9i$ (Use imaginary number.)

Example 9

Solve using the square root method.

$3x^2 = 48$

$x^2 = \dfrac{48}{3} \rightarrow x^2 = 16$ (First isolate x^2 by dividing both sides by 3.)

$\sqrt{x^2} = \pm \sqrt{16}$ (To cancel the square, take the square root of both sides.)

$x = \pm 4$

Example 10

The area of a circle is 305 cm². Find the radius using the square root method. [Use 3.1416 as an approximation for π]. Round the final answer to the nearest whole number.

$A = \pi r^2$ (Formula for area of a circle)

$305 = 3.1416\, r^2$ (Replace A and π.)

$\dfrac{305}{3.1416} = r^2$ (Isolate r^2.)

$\pm\sqrt{\dfrac{305}{3.1416}} = \sqrt{r^2}$ (To cancel the square, take the square root of both sides.)

$\pm 9.85 = r$ or $r = \mathbf{10}$ **cm**   (Measurement must be positive)

Example 11

If the difference of a number and 2 is squared, the result is 3. Find the numbers using the square root method.

$(x - 2)^2 = 3$

$\sqrt{(x-2)^2} = \pm\sqrt{3}$ (To cancel the square, take the square root of both sides.)

$x - 2 = \pm\sqrt{3}$ (Squared expression is cleared.)

$x = 2 \pm \sqrt{3}$ (Add 2 to both sides.)

Example 12

Solve using the square root method.

The formula $d = 4.9t^2$ is used for the distance d in meters that an object falls in t seconds. An object is dropped from the 48-foot Osceola clock tower (14.6 meters). How long does it take the object to hit the ground?

$d = 4.9t^2$

$14.6 = 4.9t^2$ (Replace d with 14.6 since the formula is for metric measurement.)

$\dfrac{14.6}{4.9} = t^2$ (Divide both sides by 4.9.)

$\pm\sqrt{\dfrac{14.6}{4.9}} = \sqrt{t^2}$ (To cancel the square, take the square root of both sides.)

$\pm 1.7 = t$ or **2 seconds**   (Time has to be positive)

Section 6.2 - Solving quadratic equations - square root method

 **Your turn...**

Solve using the square root method.

1. $x^2 = 121$

2. $x^2 = 20$

3. $x^2 = 7$

4. $(x + 1)^2 = 49$

5. $(x - 3)^2 = \dfrac{9}{16}$

6. $(x - 2)^2 = 9$

7. $(3x + 4)^2 = 6$

8. $x^2 = -64$

9. $5x^2 = 245$

10. The area of a circle is $500 \ cm^2$. Find the radius. [Use 3.1416 as an approximation for π]

11. If the sum of a number and 7 is squared, the result is 5. Find the number.

12. The formula $d = 4.9t^2$ is used for the distance d in meters that an object falls in t seconds. An object is dropped from the top of a 55-foot building (16.8 meters). How long does it take the object to hit the ground?

For more practice, access the online homework. See your syllabus for details.

Section 6.3 - Solving Quadratic Equations - Complete the Square

❖ Solving quadratic equations using the complete the square method.

❖ *Solving quadratic equations using the complete the square method*

The **complete the square** method is used in conjunction with the **square root** method. The square root method can be used if a quadratic equation is in the vertex format $a(x - h)^2 + k = 0$. However, if the quadratic equation is in the standard form $ax^2 + bx + c = 0$, we first need put the quadratic equation in the vertex format. To do this, we need to **complete the square**. Once the equation is in the vertex format, we will then be able to use the square root method to solve it.

Given a quadratic equation in the standard form $ax^2 + bx + c = 0$, below are the steps for the complete the square method:

1. Move c to the other side of the equation.
2. Divide both sides of the equation by the coefficient of x^2 or a if $a \neq 1$.
3. Take half of b, and square it.
4. Add the result of step 3 to both sides of the equation.
5. Rewrite the left side of the equation into a squared form $(x + h)^2$ (h is half of b), and simplify the right side.
6. Now using the square root method, take the square root of both sides, and add the " $\pm$ " on the right side.
7. Then solve for the variable x.

The variable x solved for is an x-intercept.

Example 1

Solve using the complete the square method.

$x^2 - 2x - 5 = 0$

$x^2 - 2x = 5$ (Move c to the other side of the equation or add 5 to both sides.)

$x^2 - 2x = 5$ (b is -2.)

$\left(-\frac{2}{2}\right)^2 = 1$ (Take half of b and square it.)

$x^2 - 2x + 1 = 5 + 1$ (Add half of b squared to both sides of the equation.)

$(x - 1)^2 = 6$ [Rewrite the left side of the equation into a squared form $(x + h)^2$ (h is half of b), and simplify the right side.]

$$\sqrt{(x-1)^2} = \pm\sqrt{6}$$ (Use the square root method.)

$$x - 1 = \pm\sqrt{6}$$ (Squared expression is cleared.)

$$x = 1 \pm\sqrt{6}$$ (Add 1 to both sides.)

Example 2

Solve using the complete the square method.

$$x^2 - 6x - 2 = 0$$

$$x^2 - 6x = 2$$ (Move c to the other side of the equation or add 2 to both sides.)

$$x^2 - 6x = 2$$ (b is -6.)

$$\left(-\frac{6}{2}\right)^2 = 9$$ (Take half of b and square it.)

$$x^2 - 6x + 9 = 2 + 9$$ (Add half of b squared to both sides of the equation.)

$$(x-3)^2 = 11$$ [Rewrite the left side of the equation into a squared form $(x+h)^2$ (h is half of b), and simplify the right side.]

$$\sqrt{(x-3)^2} = \pm\sqrt{11}$$ (Use the square root method.)

$$x - 3 = \pm\sqrt{11}$$ (Squared expression is cleared.)

$$x = 3 \pm\sqrt{11}$$ (Add 3 to both sides.)

Example 3

Solve using the complete the square method.

$$x^2 - 3x + 1 = 0$$

$$x^2 - 3x = -1$$ (Move c to the other side of the equation or subtract 1 from both sides.)

$$x^2 - 3x = -1$$ (b is -3.)

$$\left(-\frac{3}{2}\right)^2 = \frac{9}{4}$$ (Take half of b and square it.)

$$x^2 - 3x + \frac{9}{4} = -1 + \frac{9}{4}$$ (Add half of b squared to both sides of the equation.)

$$\left(x - \frac{3}{2}\right)^2 = \frac{5}{4}$$ [Rewrite the left side of the equation into a squared form $(x + h)^2$ (h is half of b), and simplify the right side.]

$$\sqrt{\left(x - \frac{3}{2}\right)^2} = \pm\sqrt{\frac{5}{4}}$$ (Use the square root method.)

$$x - \frac{3}{2} = \pm\frac{\sqrt{5}}{2}$$ (Squared expression is cleared.)

$$x = \frac{3}{2} \pm \frac{\sqrt{5}}{2}$$ (Add $\frac{3}{2}$ to both sides.)

Or $x = \frac{(3 \pm \sqrt{5})}{2}$

Example 4

Solve using the complete the square method.

$$2x^2 + 8x - 12 = 0$$

$2x^2 + 8x = 12$ (Move c to the other side of the equation or add 12 to both sides.)

$x^2 + 4x = 6$ (Divide both sides by 2 or by the coefficient of x^2 since $a \neq 1$.)

$x^2 + 4x = 6$ (b is 4.)

$\left(\frac{4}{2}\right)^2 = 4$ (Take half of b and square it.)

$x^2 + 4x + 4 = 6 + 4$ (Add half of b squared to both sides of the equation.)

$(x + 2)^2 = 10$ [Rewrite the left side of the equation into a squared form $(x + h)^2$ (h is half of b), and simplify the right side.]

$\sqrt{(x + 2)^2} = \pm\sqrt{10}$ (Use the square root method.)

$x + 2 = \pm\sqrt{10}$ (Squared expression is cleared.)

$x = -2 \pm \sqrt{10}$ (Subtract 2 from both sides.)

Note: this problem can be solved by first dividing both sides of the equations by 2 and proceed just like example 2.

Example 5

Solve using the complete the square method.

$3x^2 - x - 2 = 0$

$3x^2 - x = 2$ (Move c to the other side of the equation or add 2 to both sides.)

$x^2 - \frac{1}{3}x = \frac{2}{3}$ (Divide both sides by 3 or the coefficient of x^2 since $a \neq 1$.)

$x^2 - \frac{1}{3}x = \frac{2}{3}$ (b is $-\frac{1}{3}$.)

$\left(\frac{-\frac{1}{3}}{2}\right)^2 = \left(-\frac{1}{3} \cdot \frac{1}{2}\right)^2 = \left(-\frac{1}{6}\right)^2 = \frac{1}{36}$ (Take half of b and square it)

$x^2 - \frac{1}{3}x + \frac{1}{36} = \frac{2}{3} + \frac{1}{36}$ (Add half of b squared to both sides of the equation.)

$\left(x - \frac{1}{6}\right)^2 = \frac{25}{36}$ [Rewrite the left side of the equation into a squared form $(x + h)^2$ (h is half of b), and simplify the right side.]

$\sqrt{\left(x - \frac{1}{6}\right)^2} = \pm\sqrt{\frac{25}{36}}$ (Use the square root method.)

$x - \frac{1}{6} = \pm\frac{5}{6}$ (Squared expression is cleared.)

$x = \frac{1}{6} \pm \frac{5}{6}$ (Add $\frac{1}{6}$ to both sides.)

Split into 2 equations and solve.

$x = \frac{1}{6} + \frac{5}{6} = \frac{6}{6} = 1 \rightarrow x = 1$

$x = \frac{1}{6} - \frac{5}{6} = -\frac{4}{6} = -\frac{2}{3} \rightarrow x = -\frac{2}{3}$

Section 6.3 - Solving quadratic equations - complete the square

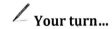 **Your turn...**

Solve using the complete the square method.

1. $x^2 - 4x - 3 = 0$

2. $x^2 - 10x - 1 = 0$

3. $x^2 - 5x + 3 = 0$

4. $4x^2 - 8x - 24 = 0$

5. $2x^2 - 3x - 2 = 0$

For more practice, access the online homework. See your syllabus for details.

Section 6.4 - Solving Quadratic Equations - Quadratic Formula

❖ Solving quadratic equations using the quadratic formula

❖ *Solving quadratic equations using the quadratic formula*

When solving an equation using the quadratic formula, first the **equation needs to be set equal to zero.** An equation in the form of $ax^2 + bx = -c$ or $ax^2 = -bx - c$ is not ready. It needs to be in the form $ax^2 + bx + c = 0$. Then identify a, b, and c and replace them in the quadratic formula below and solve. *The solutions or the x-values found are actually x-intercepts.*

$$x = \frac{-b \pm \sqrt{b^2 - 4ac}}{2a}$$

Example 1

Solve using the quadratic formula.

$3x^2 - 8x + 4 = 0$

$a = 3, \qquad b = -8, \qquad c = 4$

$$x = \frac{-(-8) \pm \sqrt{(-8)^2 - 4(3)(4)}}{2(3)}$$ (Replace a, b, c in the formula.)

$$x = \frac{8 \pm \sqrt{16}}{6} = \frac{8 \pm 4}{6}$$

Split into 2 equations and solve.

$$x = \frac{8 + 4}{6} = \frac{12}{6} = 2 \rightarrow x = 2$$

$$x = \frac{8 - 4}{6} = \frac{4}{6} = \frac{2}{3} \rightarrow x = \frac{2}{3}$$

Example 2

Solve using the quadratic formula.

$x^2 - x - 5 = 0$

$a = 1, \qquad b = -1, \qquad c = -5$

$$x = \frac{-(-1) \pm \sqrt{(-1)^2 - 4(1)(-5)}}{2(1)} \rightarrow x = \frac{1 \pm \sqrt{21}}{2}$$ (Replace a, b, c in the formula and solve.)

Example 3

Solve using the quadratic formula.

$$2x^2 + 8x + 7 = 0$$

$$a = 2, \qquad b = 8, \qquad c = 7$$

$$x = \frac{-(8) \pm \sqrt{(8)^2 - 4(2)(7)}}{2(2)}$$ (Replace a, b, c in the formula.)

$$x = \frac{-8 \pm \sqrt{8}}{4}$$

$$x = \frac{-8 \pm \sqrt{4 \cdot 2}}{4} = \frac{-8 \pm 2\sqrt{2}}{4}$$ (Simplify.)

$$x = \frac{-4 \pm \sqrt{2}}{2}$$ (Reduce all **numbers outside** of the radical sign.)

Example 4

Solve using the quadratic formula.

$$4x^2 + x + 1 = 0$$

$$a = 4, \qquad b = 1, \qquad c = 1$$

$$x = \frac{-(1) \pm \sqrt{(1)^2 - 4(4)(1)}}{2(4)}$$ (Replace a, b, c in the formula.)

$$x = \frac{-(1) \pm \sqrt{-15}}{8} = \frac{-1 \pm \sqrt{-1}\sqrt{15}}{8} = \frac{-1 \pm i\sqrt{15}}{8}$$

Example 5

Solve using the quadratic formula.

The area of a rectangular flyer advertising the Osceola math fair is 22 in^2. The length of the flyer is 2 inches more than the width. What are the length and the width of the flyer?

$A = 22$

$L = w + 2$

$A = L \cdot w$ (Area of a rectangle)

Replace A with 22 and L with $w + 2$ in the area formula.

$22 = (w + 2) \cdot w \rightarrow 22 = w^2 + 2w$

$w^2 + 2w - 22 = 0$ (Subtract 22 from both sides.)

This equation cannot be factored. Use the quadratic formula to solve it.

$w^2 + 2w - 22 = 0$

$a = 1, \quad b = 2, \quad c = -22$

$$w = \frac{-(2) \pm \sqrt{(2)^2 - 4(1)(-22)}}{2(1)}$$ (Replace a, b, c in the formula.)

$$w = \frac{-(2) \pm \sqrt{92}}{2} = \frac{-2 \pm 9.59}{2}$$

Split into 2 equations and solve.

$$w = \frac{-2 + 9.59}{2} = 3.795 \text{ or } \mathbf{w = 3.8 \text{ inches}}$$

$$w = \frac{-2 - 9.59}{2} = -5.795 \text{ or } -5.8$$ (Ignore $-$ negative measurement)

Now find the length.

$L = w + 2$

$L = 3.8 + 2 \rightarrow l = \mathbf{5.8 \text{ inches}}$

Example 6

Solve using the quadratic formula.

After the discussion of quadratic functions in class, Madeline is eager to practice what she learns. She goes to the last floor of building four of the Osceola campus and throws an object up in the air. The function that models this scenario is given by $h(t) = -16t^2 + 32t + 92$, where $h(t)$ is the height of the object above the ground and t is the time in seconds. She wants to find out when will the object hit the ground?

When the object hits the ground, the height is zero. Replace $h(t)$ with 0.

$h(t) = -16t^2 + 32t + 92$

$0 = -16t^2 + 32t + 92$

$\frac{0}{-4} = \frac{-16}{-4}t^2 + \frac{32}{-4}t + \frac{92}{-4}$ (Simplify.)

$0 = 4t^2 - 8t - 23$ (This equation cannot be factored.)

Use the quadratic formula.

$4t^2 - 8t - 23 = 0$

$a = 4, \qquad b = -8, \qquad c = -23$

$t = \dfrac{-(-8) \pm \sqrt{(-8)^2 - 4(4)(-23)}}{2(4)}$ (Replace a, b, c in the formula.)

$t = \dfrac{8 \pm \sqrt{64 + 368}}{8} = \dfrac{8 \pm \sqrt{432}}{8}$

$t = \dfrac{8 \pm 20.78}{8}$

Split into 2 equations and solve.

$t = \dfrac{8 + 20.78}{8} = 3.597$ or **3.6 seconds**

$t = \dfrac{8 - 20.78}{8} = -1.598$ (Ignore − negative time)

It will take approximately **3.6 seconds** for the object to hit the ground.

Example 7

Solve using the quadratic formula.

Intramural sports at Valencia College give students the opportunity to be active. On Osceola campus, soccer is one of the intramural sports available to students. During one of the practice sessions, one of the students, Sienna, kicked the soccer ball straight up from the ground with an initial velocity of 48 feet per second. Its height above the ground in feet is given by $h(t) = -16t^2 + 48t$. When will the ball reach a height of 24 ft.?

Replace $h(t)$ with 24.

$$h(t) = -16t^2 + 48t$$

$$24 = -16t^2 + 48t$$

$$0 = -16t^2 + 48t - 24 \qquad \text{(Subtract 24 from both sides to set the function equal to 0.)}$$

$$\frac{0}{-8} = \frac{-16}{-8}t^2 + \frac{48}{-8}t - \frac{24}{-8} \qquad \text{(Simplify.)}$$

$$0 = 2t^2 - 6t + 3 \qquad \text{(This equation cannot be factored.)}$$

Use the quadratic formula.

$$2t^2 - 6t + 3 = 0$$

$$a = 2, \qquad b = -6, \qquad c = 3$$

$$t = \frac{-(-6) \pm \sqrt{(-6)^2 - 4(2)(3)}}{2(2)} \qquad \text{(Replace a, b, c in the formula.)}$$

$$t = \frac{6 \pm \sqrt{36 - 24}}{4} = \frac{6 \pm \sqrt{12}}{4}$$

$$t = \frac{6 \pm 3.46}{4}$$

Split into 2 equations and solve.

$$t = \frac{6 + 3.46}{4} = \textbf{2.37 seconds}$$

$$t = \frac{6 - 3.46}{4} = \textbf{0.64 second}$$

The ball will reach a height of 24 ft. at $t = 0.64$ second and then $t = 2.37$ seconds (on the way up and when coming back down).

Example 8

Solve using the quadratic formula.

The product of two consecutive positive even integers is 624. Find the two integers.

Let x be the first positive even integer.
Let $x + 2$ be the second even positive integer. The product of the two is $(x)(x + 2)$.

The equation becomes:

$$x(x + 2) = 624$$

$$x^2 + 2x = 624 \qquad \text{(Distribute the } x.)$$

$$x^2 + 2x - 624 = 0 \qquad \text{(Subtract 624 from both sides.)}$$

Use the quadratic formula.

$$x^2 + 2x - 624 = 0$$

$$a = 1, \qquad b = 2, \qquad c = -624$$

$$x = \frac{-(2) \pm \sqrt{(2)^2 - 4(1)(-624)}}{2(1)} \qquad \text{(Replace a, b, c in the formula.)}$$

$$x = \frac{-2 \pm \sqrt{4 + 2496}}{2} = \frac{-2 \pm \sqrt{2500}}{2}$$

$$x = \frac{-2 \pm 50}{2}$$

Split into 2 equations and solve.

$$x = \frac{-2 + 50}{2} = \mathbf{24}$$

$$x = \frac{-2 - 50}{2} = \mathbf{-26}$$

Discard the negative answer.

Remember $x + 2$ is the second even positive integer.

$$x + 2 = 24 + 2 = \mathbf{26}$$

The two consecutive positive even integers are: **24 and 26.**

Section 6.4 - Solving quadratic equations - quadratic formula

✎ **Your turn...**

Solve using the quadratic formula.

1. $x^2 - x - 3 = 0$

2. $2x^2 + 8x + 5 = 0$

3. $2x^2 - 5x + 2 = 0$

4. $7x^2 + x + 2 = 0$

5. $x^2 - x - 1 = 0$

6. $3x^2 + 5x + 2 = 0$

7. The area of a rectangular flyer advertising the Osceola math fair is 25 in². The length of the flyer is 3 inches more than the width. What are the length and the width of the flyer?

8. After the discussion of quadratic functions in class, Patricia is eager to practice what she learns. She goes to the rooftop of a very high building and throws an object in the air. The function that models this scenario is given by $h(t) = -16t^2 + 32t + 80$, where $h(t)$ is the height of the object above the ground and t is the time in seconds. She wants to find out when will the object hit the ground?

9. Intramural sports at Valencia College give students the opportunity to be active. On Osceola campus, soccer is one of the intramural sports available to students. During one of the practice sessions, one of the students, Akiyoshi, kicked the soccer ball straight up from the ground with an initial velocity of 56 feet per second. Its height above the ground in feet is given by $h(t) = -16t^2 + 56t$. When will the ball reach a height of 32 ft.?

10. The product of two consecutive positive even integers is 1088. Find the two integers.

For more practice, access the online homework. See your syllabus for details.

Section 6.5 - Graphing Quadratic Functions

- ❖ Introduction to the graph of quadratic functions
- ❖ Vertical shift
- ❖ Horizontal shift
- ❖ Shape of the graph (wide and narrow)
- ❖ Combination of vertical and horizontal shifts

❖ *Introduction to the graph of quadratic functions*

The graph of a quadratic function $f(x) = ax^2 + bx + c$ is called a parabola. The parabola either opens up or down based on the coefficient of x^2 or a. If the coefficient is positive, the graph opens up. If the coefficient is negative, the graph opens down.

If the graph opens up, then the lowest point is called vertex. If the graph opens down, then the highest point is called vertex. In summary, the vertex is either a minimum point (lowest) or a maximum point (highest) based on the shape of the graph.

One more feature of the parabola is that it is symmetric with respect to an axis or axis of symmetry. The axis is always the first coordinate of the vertex (h, k) or an equation of a vertical line expressed as $x = h$. It can be described as the vertical line that goes through the vertex.

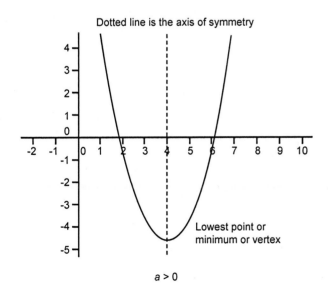

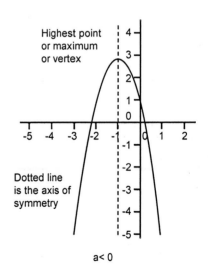

❖ *Vertical shift*

A vertical shift can be described as the movement of the graph vertically or along the y-axis. It can be expressed as $f(x) + k$ or $f(x) - k$, where k is the vertical shift of the function.

The constant k determines whether the graph is shifted up or down. If k is positive, then the graph is shifted k units up. If k is negative, then the graph is shifted k units down.

Example 1

$f(x) = x^2 + 1$

Give a written description for the shift. Graph the function.

The quadratic function or square function , $f(x) = x^2$, is shifted up 1 unit.

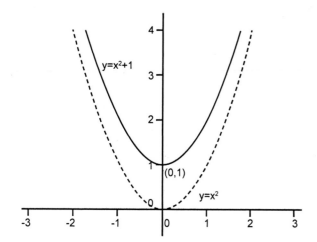

The vertex is $(0,1)$ and the axis of symmetry is $x = 0$.

Example 2

$f(x) = x^2 - 3$

Give a written description for the shift. Graph the function.

The quadratic function or square function , $f(x) = x^2$, is shifted down 3 units.

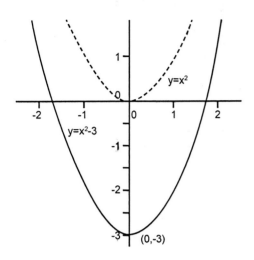

The vertex is $(0, -3)$ and the axis of symmetry is $x = 0$.

Example 3

$$f(x) = x^2 - \frac{1}{2}$$

Give a written description for the shift. Graph the function.

The quadratic function or square function , $f(x) = x^2$, is shifted down $\frac{1}{2}$ unit.

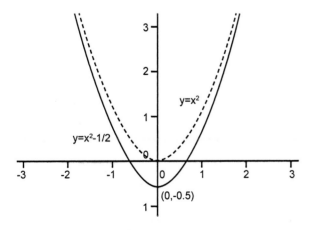

The vertex is $(0, -\frac{1}{2})$ and the axis of symmetry is $x = 0$.

Example 4

$$f(x) = -x^2 + 3$$

Give a written description for the shift. Graph the function.

The quadratic function or square function, $f(x) = -x^2$, is shifted up 3 units. It opens down.

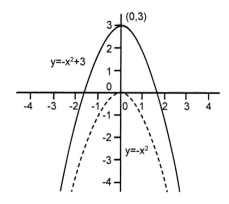

The vertex is $(0, 3)$ and the axis of symmetry is $x = 0$.

❖ *Horizontal shift*

A horizontal shift can be described as the movement of the graph horizontally or along the x-axis. It can be expressed as $f(x + k)$ or $f(x - k)$.

The constant k determines whether the graph is shifted left or right. If k is positive, then the graph is shifted k units to the left. If k is negative, then the graph is shifted k units to the right. The graph will move in the opposite direction of k.

Example 5

$$f(x) = (x + 4)^2$$

Give a written description for the shift. Graph the function.

The quadratic function or square function, $f(x) = x^2$, is shifted 4 units to the left.

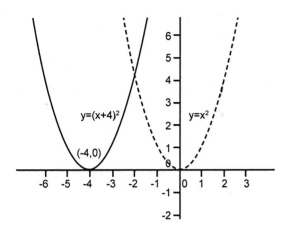

The vertex is $(-4, 0)$ and the axis of symmetry is $x = -4$.

Example 6

$$f(x) = (x - 2)^2$$

Give a written description for the shift. Graph the function.

The quadratic function or square function, $f(x) = x^2$, is shifted 2 units to the right.

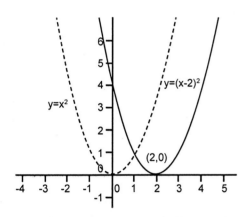

The vertex is $(2, 0)$ and the axis of symmetry is $x = 2$.

Example 7

$$f(x) = \left(x - \frac{3}{4}\right)^2$$

Give a written description for the shift. Graph the function.

The quadratic function or square function, $f(x) = x^2$, is shifted $\frac{3}{4}$ of a unit to the right.

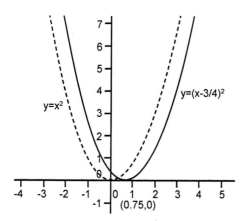

The vertex is $(\frac{3}{4}, 0)$ and the axis of symmetry is $x = \frac{3}{4}$.

Example 8

$f(x) = -(x + 2)^2$

Give a written description for the shift. Graph the function.

The quadratic function or square function, $f(x) = -x^2$, is shifted 2 units to the left. It opens down.

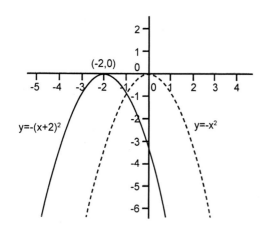

The vertex is $(-2, 0)$ and the axis of symmetry is $x = -2$.

❖ *Shape of the Graph (wide and narrow)*

The coefficient, a, of $f(x) = ax^2$ determines whether the graph is narrow or wide. If $|a| > 1$, then the graph is narrower than the original function $f(x) = x^2$. If $|a| < 1$ (fraction or decimal), then the graph is wider that the original function $f(x) = x^2$.

Example 9

Graph the function.

$f(x) = 2x^2$

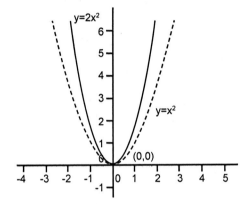

Because the coefficient 2 is greater than 1, the graph is narrower than the basic function $f(x) = x^2$.

The vertex is $(0, 0)$ and the axis of symmetry is $x = 0$.

Example 10

Graph the function $f(x) = \frac{1}{2}x^2$.

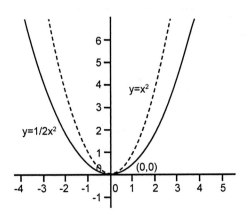

Because the coefficient $\frac{1}{2}$ is less than 1, the graph is wider than the basic function $f(x) = x^2$.

The vertex is $(0, 0)$ and the axis of symmetry is $x = 0$.

Example 11

Graph the function $f(x) = -\frac{1}{3}x^2$.

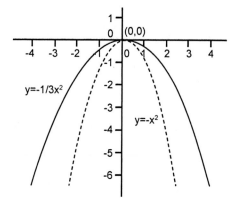

Because the coefficient $\frac{1}{3}$ is less than 1, the graph is wider than the function $f(x) = -x^2$.

The negative sign means that the graph opens down.

The vertex is $(0, 0)$ and the axis of symmetry is $x = 0$.

❖ *Combination of vertical and horizontal shifts*

A combined shift can be described as the movement of the graph horizontally or along the x-axis and then vertically or along the y-axis. It is in the vertex format $f(x) = a(x - h)^2 + k$. The vertex is (h, k) and the axis of symmetry is $x = h$.

Example 12

$f(x) = (x - 2)^2 + 3$

Give a written description for the shift. Graph the function.

The quadratic function or square function, $f(x) = x^2$, is shifted 2 units to the right and up 3 units.

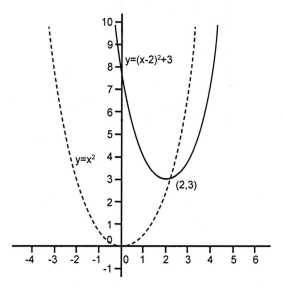

The vertex is $(2, 3)$ and the axis of symmetry is $x = 2$. The graph opens up because the coefficient in front of the parenthesis is positive.

Example 13

$$f(x) = (x + 3)^2 - 4$$

Give a written description for the shift. Graph the function.

The quadratic function or square function, $f(x) = x^2$, is shifted 3 units to the left and down 4 units.

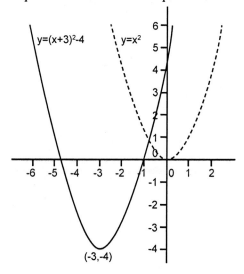

The vertex is $(-3, -4)$ and the axis of symmetry is $x = -3$. The graph opens up because the coefficient in front of the parenthesis is positive.

Example 14

$$f(x) = (x-1)^2 - 2$$

Give a written description for the shift. Graph the function.

The quadratic function or square function, $f(x) = x^2$, is shifted 1 unit to the right and down 2 units.

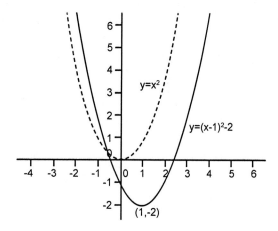

The vertex is $(1, -2)$ and the axis of symmetry is $x = 1$. The graph opens up because the coefficient in front of the parenthesis is positive.

Example 15

$$f(x) = (x+5)^2 + 4$$

Give a written description for the shift. Graph the function.

The quadratic function or square function, $f(x) = x^2$, is shifted 5 units to the left and up 4 units.

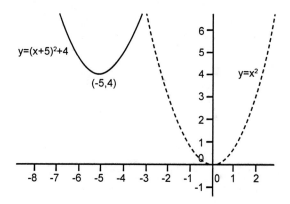

The vertex is $(-5, 4)$ and the axis of symmetry is $x = -5$. The graph opens up because the coefficient in front of the parenthesis is positive.

Example 16

$$f(x) = -(x + 2)^2 + 4$$

Give a written description for the shift. Graph the function.

The quadartic function or square funciton, $f(x) = -x^2$, is shifted 2 units to the left and up 4 units.

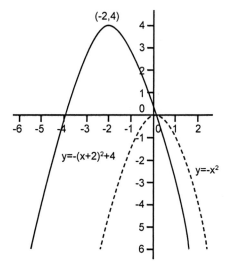

The vertex is $(-2, 4)$ and the axis of symmetry is $x = -2$. The graph opens down because the coefficient in front of the parenthesis is negative.

Section 6.5 – Graphing quadratic functions

Your turn...

Graph the quadratic functions.

1. $f(x) = x^2 + 2$

2. $f(x) = x^2 - 1$

3. $f(x) = x^2 - \frac{1}{4}$

4. $f(x) = -x^2 + 3$

5. $f(x) = (x + 1)^2$

6. $f(x) = (x - 2)^2$

7. $f(x) = -(x + 5)^2$

8. $f(x) = 4x^2$

9. $f(x) = -\frac{1}{2}x^2$

10. $f(x) = (x - 1)^2 + 2$

11. $f(x) = (x + 4)^2 - 3$

12. $f(x) = (x - 5)^2 - 2$

13. $f(x) = (x + 1)^2 + 4$

14. $f(x) = -(x + 3)^2 + 6$

For more practice, access the online homework. See your syllabus for details.

Section 6.6 - Finding the Vertex of Quadratic Functions

- ❖ Finding the vertex
  - o By inspection
  - o Using the vertex formula
  - o Using a graphing calculator
- ❖ Applications

Let us review the vertex again. So far, we looked at the graph and identified the point that is the minimum or maximum. We will discuss three ways of finding the vertex.

- ❖ *Finding the vertex*

  - o *By inspection*

If the quadratic equation is in the format $f(x) = a(x - h)^2 + k$, then the vertex is (h, k).

Example 1

Find the vertex of the quadratic function.

$$f(x) = -(x - 2)^2 + 3$$

h is 2 and k is 3.

The vertex is $(2, 3)$.

Example 2

Find the vertex of the quadratic function.

$$f(x) = (x + 3)^2 - 4$$

$$f(x) = (x - (-3))^2 - 4$$

h is -3 and k is -4.

The vertex is $(-3, -4)$.

Example 3

Find the vertex of the quadratic function.

$$f(x) = (x - 1)^2 - 2$$

h is 1 and k is -2.

The vertex is $(1, -2)$.

Example 4

Find the vertex of the quadratic function.

$$f(x) = (x + 5)^2 + 4$$

$$f(x) = (x - (-5))^2 + 4$$

h is -5 and k is 4.

The vertex is $(-5, 4)$.

○ *Using the vertex formula*

Another way to find the vertex is to use the vertex formula. If the equation is in the format $f(x) = ax^2 + bx + c$, then the vertex is (x, y). Solve for x by using the formula $x = -\frac{b}{2a}$. Then plug in x into the function $f(x)$ to get y.

Example 5

Find the vertex of the quadratic function.

$$f(x) = x^2 - 4x + 4$$

$a = 1$ and $b = -4$

$$x = -\frac{b}{2a} = -\frac{-4}{2(1)} = \frac{4}{2} = 2$$

$$y = (2)^2 - 4(2) + 4 = 4 - 8 + 4 = 0 \qquad \text{[Plug in } x \text{ into } f(x).]$$

The vertex is $(2, 0)$.

Example 6

Find the vertex of the quadratic function.

$$f(x) = 2x^2 - 4x + 5$$

$a = 2$ and $b = -4$

$$x = -\frac{b}{2a} = -\frac{(-4)}{2(2)} = \frac{4}{4} = 1$$

$$y = 2(1)^2 - 4(1) + 5 = 2 - 4 + 5 = 3 \qquad \text{[Plug in } x \text{ into } f(x).]$$

The vertex is $(1, 3)$.

Example 7

Find the vertex of the quadratic function.

$$f(x) = 4x^2 + 16x + 13$$

$a = 4$ and $b = 16$

$$x = -\frac{b}{2a} = -\frac{16}{2(4)} = \frac{-16}{8} = -2$$

$$y = 4(-2)^2 + 16(-2) + 13 = 16 - 32 + 13 = -3 \qquad \text{[Plug in } x \text{ into } f(x).]$$

The vertex is $(-2, -3)$.

○ *Using a graphing calculator (TI 83 or 84)*

Graph the function, and using the following steps, find the vertex.

1) If the graph of the quadratic function or **parabola opens up, it has a minimum**.
   a. Use 2ND trace, press 3 for minimum. The prompt "left bound?" is displayed.
   b. Move the cursor to the left of the lowest point and press enter.
   c. The next prompt "right bound?" is displayed.
   d. Move the cursor to the right of the lowest point and press enter.
   e. The last prompt "Guess?" is displayed. Press enter.

   The minimum or vertex is displayed at the bottom of the calculator screen.

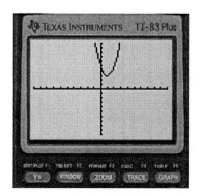  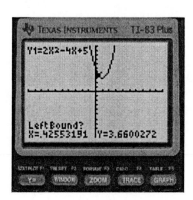

2) If the graph of the quadratic function or **parabola opens down, it has a maximum**.
   a. Use 2ND trace, press 4 for maximum. The prompt "left bound?" is displayed.
   b. Move the cursor to the left of the highest point and press enter.
   c. The next prompt "right bound?" is displayed.
   d. Move the cursor to the right of the highest point and press enter.
   e. The last prompt "Guess?" is displayed. Press enter.

   The maximum or vertex is displayed at the bottom of the calculator screen.

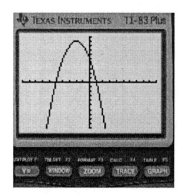 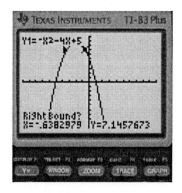 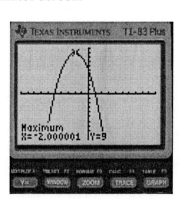

Example 8

Use a graphing calculatro to graph the function and to find the vertex. (*See page 256 for instructions*)

$$f(x) = -x^2 + 2x + 3$$

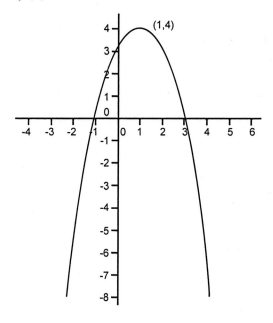

The vertex is $(1, 4)$.

Example 9

Use a graphing calculatro to graph the function and to find the vertex. (*See page 256 for instructions*)

$$f(x) = 2x^2 + 8x + 3$$

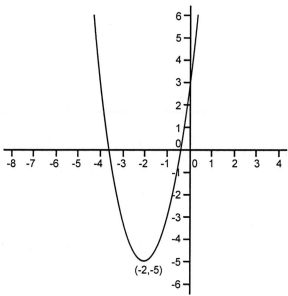

The vertex is $(-2, -5)$.

❖ *Applications*

When solving applications that require the use of the vertex formula, one of the following key words will be mentioned: maximum, minimum, least, most, highest, lowest, or any other synonyms of these words. In other words, if any of the previously mentioned words is stated in the application, then use the vertex formula (-b/2a).

Example 10

Solve the application.

The function for an object that is thrown up in the air is given by $h(t) = -16t^2 + 48t + 64$, where $h(t)$ is the height of the object above the ground and t is the time in seconds. When will the object reach the maximum height? What is the maximum height of the object?

$h(t) = -16t^2 + 48t + 64$

$a = -16$ and $b = 48$

Find the time using the vertex formula.

$$t = -\frac{b}{2a} = -\frac{48}{2(-16)} = \frac{-48}{-32} = 1.5$$

The object will reach the maximum height after **1.5 seconds**.

To find the maximum height, replace t with 1.5 in the height equation.

$h(1) = -16(1.5)^2 + 48(1.5) + 64 = -36 + 72 + 64 = 100$

The object will reach a maximum height of **100 ft**.

Example 11

Solve the application.

Jade wants to sell graphing calculators. The revenue function is given by $R(p) = -15p^2 + 4500p$, where $R(p)$ is the revenue and p is the price in dollars. What is the maximum revenue?

$R(p) = -15p^2 + 4500p$

Find the price of a calculator using the vertex formula.

$a = -15$ and $b = 4500$

$$p = -\frac{b}{2a} = -\frac{4500}{2(-15)} = \frac{-4500}{-30} = \$150$$

Plug in 150 into the revenue function $R(p)$.

$R = -15(150)^2 + 4500(150) = -337,500 + 675,000 = \$337,500$

The maximum revenue is **$337,500**.

Section 6.6 – Finding the vertex of quadratic functions

 **Your turn…**

Find the vertex by inspection.

1. $f(x) = -(x-8)^2 + 3$
2. $f(x) = (x+1)^2 - 4$
3. $f(x) = (x-5)^2 - 2$
4. $f(x) = (x+7)^2 + 4$

Find the vertex using the vertex formula.

5. $f(x) = x^2 - 8x + 4$
6. $f(x) = 2x^2 - 12x + 5$
7. $f(x) = 4x^2 + 24x + 13$

Use a graphing calculator to graph the function and to find the vertex. *(See page 256 for instructions)*

8. $f(x) = -x^2 + 4x + 3$
9. $f(x) = 2x^2 + 8x + 5$

Applications of the vertex formula

10. The function for an object that is thrown up in the air is given by $h(t) = -16t^2 + 64t + 94$, where $h(t)$ is the height of the object above the ground and t is the time in seconds. What is the maximum height of the object?

11. Chang wants to sell graphing calculators. The revenue function is given by $R(p) = -10p^2 + 3500p$, where $R(p)$ is the revenue and p is the price in dollars. What is the maximum revenue?

12. The Phi Theta Kappa officers plan to raise money for a trip. They set up a cookie stand in the atrium of building two of the Osceola campus for 30 days. The daily cost of operating this cookie stand is modeled by the function, $C(x) = x^2 - 36x + 500$, where $C(x)$ is the cost in dollars and x is the number cookies sold. a) Find the number of cookies that must be sold to minimize the cost. b) What is the minimum cost?

For more practice, access the online homework. See your syllabus for details.

Chapter 6: Review of Terms, Concepts, and Formulas

- When solving an equation by **factoring**, first make sure the equation is **equal to zero**, and then factor the equation using the techniques learned in section 3.5. Then set each factor equal to zero (use the zero-product rule) and solve.

- When solving an equation using the **square root method**, make sure that the expression with the square root is isolated, and then take the square root of both sides. Add $\pm$ in front of the radical sign on the right side. The power is two, so there should be two possible answers. Because the left side has the same index and power, the radical will go away or turns to 1. Then solve the equation by isolating the variable.

- Given a quadratic equation in the standard form $ax^2 + bx + c = 0$, below are the steps for the **complete the square method**:
  - Move c to the other side of the equation.
  - Divide both sides of the equation by the coefficient of x^2 or a if $a \neq 1$.
  - Take half of b, and square it.
  - Add the result of step 3 to both sides of the equation.
  - Rewrite the left side of the equation into a squared form $(x + h)^2$ (h is half of b), and simplify the right side.
  - Now using the square root method, take the square root of both sides, and add the " $\pm$ " on the right side.
  - Then solve for the variable x.

- When solving an equation using the **quadratic formula**, first the equation needs to be set equal to zero. An equation in the form of $ax^2 + bx = -c$ or $ax^2 = -bx - c$ is not ready. It needs to be in the form $ax^2 + bx + c = 0$. Then identify a, b, and c and replace them in the quadratic formula below and solve.

$$x = \frac{-b \pm \sqrt{b^2 - 4ac}}{2a}$$

- The **graph of a quadratic function** $f(x) = ax^2 + bx + c$ is called a parabola. The parabola either opens up or down based on the coefficient of x^2 or a. If the coefficient is positive, the graph opens up. If the coefficient is negative, the graph opens down.

- If the graph opens up, then the lowest point is called **vertex**. If the graph opens down, then the highest point is called **vertex**. In summary, the vertex is either a minimum point (lowest) or a maximum point (highest) based on the shape of the graph.

- One more feature of the parabola is that it is symmetric with respect to an axis or **axis of symmetry**. The axis is always the first coordinate of the vertex (h, k) or an equation of a vertical line expressed as $x = h$. It can be described as the vertical line that goes through the vertex.

- A **vertical shift** can be described as the movement of the graph vertically or along the y-axis. It can be expressed as $f(x) + k$ or $f(x) - k$, where k is the vertical shift of the function. The constant k determines whether the graph is shifted up or down. If k is positive, then the graph is shifted k units up. If k is negative, then the graph is shifted k units down.

- A **horizontal shift** can be described as the movement of the graph horizontally or along the x-axis. It can be expressed as $f(x + k)$ or $f(x - k)$. The constant k determines whether the graph is shifted left or right. If k is positive, then the graph is shifted k units to the left. If k is negative, then the graph is shifted k units to the right. The graph will move in the opposite direction of k.

- The coefficient of x^2 determines whether the graph is **narrow or wide**. If the coefficient is a whole number greater than 1, then the graph is narrower than the original function $f(x) = x^2$. If the coefficient is less than 1, then the graph is wider that the original function.

- To find the vertex **by inspection**, look at the equation. If the quadratic equation is in the format $f(x) = a(x - h)^2 + k$, then the vertex is (h, k).

- To find the vertex using the **vertex formula** - If the equation is in the format $f(x) = ax^2 + bx + c$, then the vertex is (x, y). Solve for x by using the formula $x = -\dfrac{b}{2a}$. Then plug in x into the function $f(x)$ to get y.

- To find the vertex using a **graphing calculator** (TI 83 or 84):

If the graph of the quadratic function or **parabola opens up, it has a minimum**.
  - Use 2ND trace, press 3 for minimum. The prompt "left bound?" is displayed.
  - Move the cursor to the left of the lowest point and press enter.
  - The next prompt "right bound?" is displayed.
  - Move the cursor to the right of the lowest point and press enter.
  - The last prompt "Guess?" is displayed. Press enter.

If the graph of the quadratic function or **parabola opens down, it has a maximum**.
  - Use 2ND trace, press 4 for maximum. The prompt "left bound?" is displayed.
  - Move the cursor to the left of the highest point and press enter.
  - The next prompt "right bound?" is displayed.
  - Move the cursor to the right of the highest point and press enter.
  - The last prompt "Guess?" is displayed. Press enter.

Appendix A - Linear Functions

❖ Data Analysis of linear functions

❖ Data Analysis of linear functions

A linear function is a function, which graph is a straight line. A linear function can have the following form, $y = f(x) = mx + b$. A linear function has one independent variable and one dependent variable. The independent variable is x and the dependent variable is y. The letter, b, is the constant term or the y-intercept. It is the value of the dependent variable when $x = 0$. The letter, m, is the coefficient of the independent variable. It is also known as the slope.

When analyzing a set of given data, use the data to find the slope. If the slope is constant or the same, then the data can be modeled by a linear function.

Example 1

Decide if the data can be modeled by a linear function. If yes, find the linear function.

x	Y or f(x)
0	-10
2	-6
4	-2
6	2
8	6

First, use the points in the above table to calculate the slope. If the slope is constant or the same, then the data can be modeled by a linear function.

Using $(0, -10)$ and $(2, -6)$, find the slope.

$$m = \frac{y_2 - y_1}{x_2 - x_1} = \frac{-6 - (-10)}{2 - 0} = \frac{4}{2} = 2$$

Repeat the process using the following points: $(2, -6)$ and $(4, -2)$; $(4, -2)$ and $(6, 2)$; $(6, 2)$ and $(8, 6)$. Once the slope is found for all the points, the result is 2. The slope is constant, thus the data can be modeled by a linear function. Now find the linear function using $f(x) = mx + b$.

We have m or the slope and it is 2. We also have b, which is the y-intercept -10. The linear function for the above set of data is $f(x) = 2x - 10$.

Example 2

Decide if the data can be modeled by a linear function. If yes, find the linear function.

x	Y or f(x)
1	-1
4	-10
7	-19
10	-28
13	-37

First, use the points in the above table to calculate the slope. If the slope is constant or the same, then the data can be modeled by a linear function.

Using $(1, -1)$ and $(4, -10)$, find the slope.

$$m = \frac{y_2 - y_1}{x_2 - x_1} = \frac{-10 - (-1)}{4 - 1} = \frac{-9}{3} = -3$$

Repeat the process using the following points: $(4, -10)$ and $(7, -19)$; $(7, -19)$ and $(10, -28)$; and lastly

$(10, -28)$ and $(13, -37)$. Once the slope is found for all the points, the result is -3. The slope is constant, thus the data can be modeled by a linear function. Now find the linear function using $f(x) = mx + b$.

We have m or the slope and it is -3. By looking at the table, we do not have b, which is the y-intercept. Let us find b using one of the points and replace the coordinates of the points and the slope in this function, $f(x) = mx + b$.

Using $(7, -19)$, replace x with $7, y$ or $f(x)$ with -19, and m with -3,

$f(x) = mx + b$

$-19 = -3(7) + b$

$-19 = -21 + b$

$-19 + 21 = b$

$\qquad 2 = b$

The linear function for the above set of data is $f(x) = -3x + 2$.

Example 3 – Decide if the data can be modeled by a linear function. If yes, find the linear function.

x	Y or f(x)
1	3
2	6
3	7
4	20
5	24

First, use the points in the above table to calculate the slope. If the slope is constant or the same, then the data can be modeled by a linear function.

Using $(1, 3)$ and $(2, 6)$, find the slope.

$$m = \frac{y_2 - y_1}{x_2 - x_1} = \frac{6 - 3}{2 - 1} = \frac{3}{2}$$

Repeat the process using the following points: $(2, 6)$ and $(3, 7)$; $(3, 7)$ and $(4, 20)$; and lastly $(4, 20)$ and $(5, 24)$. Once the slope is found for all the points, the result is not the same. The slope is not constant, thus the data cannot be modeled by a linear function.

✎ **Your turn...**

Decide if the set of data below can be modeled by a linear function. If yes, find the linear function.

x	Y or f(x)
0	-7
2	3
4	13
6	23
8	33

1.

x	Y or f(x)
1	1/2
2	1
3	2
4	5
5	7

2.

x	Y or f(x)
0	2
3	1
6	0
9	-1

3.

x	Y or f(x)
5	-1.25
10	2.5
15	6.25
20	10

4.

x	Y or f(x)
2	-15
4	-31
6	-47
8	-63

5.

x	Y or f(x)
-2	1
-1	8
0	10
1	11
2	14

6.

Answers: 1. $f(x) = 5x - 7$ 2. Not linear   3. $f(x) = -\frac{1}{3}x + 2$

4. $f(x) = \frac{3}{4}x - 5$ 5. $f(x) = -8x + 1$ 6. Not linear

---

Appendix B

Section 2.5 – Graphing Linear Inequalities and Systems of Linear Inequalities

- ❖ Graphing linear inequalities – <u>By hand</u>
- ❖ Graphing systems of linear inequalities – <u>By hand</u>

❖ *Graphing linear inequalities- By hand*

Example 1

Graph the linear inequality $y > 4x + 4$.

First graph the equation $y = 4x + 4$.

Make a table of values or use intercepts to graph the line.

Let us use intercepts.

$x = 0 \rightarrow \quad y = 4(0) + 4 \rightarrow \quad y = 4$; The y-intercept is $(0, 4)$.

$y = 0 \rightarrow \quad 0 = 4x + 4 \rightarrow \quad -4 = 4x \rightarrow x = -1$; The x-intercept is $(-1, 0)$.

Use the points $(0, 4)$ and $(-1, 0)$ to draw the line.

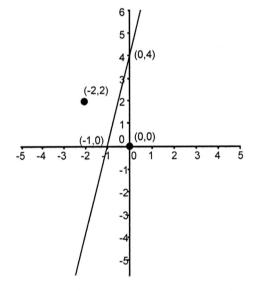

Next, find the solutions or the points that make the inequality true. Then shade the area where the solutions are. In order to find the points or solutions that make the inequality true, choose a **couple of points**.

First, choose the point $(0, 0)$ on one side of the line and replace it in the inequality:

$0 > 4(0) + 4 \quad \rightarrow 0 > 4$. This is a false statement so $(0, 0)$ will not be in the shaded area of the graph.

Next, choose the point $(-2, 2)$ on the other side of the line and replace it in the inequality:

$2 > 4(-2) + 4 \rightarrow 2 > -4$. This is a **true statement** so $(-2, 2)$ will be the shaded area of the graph.

Shade the side of the graph where the point $(-2, 2)$ is located. If the symbol from the given inequality is $<$ or $>$, the line is a dotted line because there is no equal sign or the points on the line are not solutions or not included.

Here is the final graph of the inequality $y > 4x + 4$:

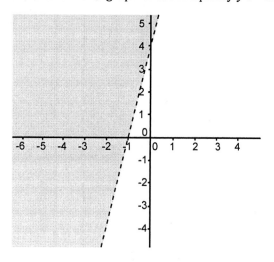

Example 2

Graph the system of linear inequality $y > 4x + 4$ and $y \le x - 1$.

First graph the equation $y = 4x + 4$.

Make a table of values or use intercepts to graph the line.

Let us use intercepts.

$x = 0 \rightarrow \quad y = 4(0) + 4 \rightarrow \quad y = 4$; The y-intercept is $(0, 4)$.

$y = 0 \rightarrow \quad 0 = 4x + 4 \rightarrow \quad -4 = 4x \rightarrow x = -1$; The x-intercept is $(-1, 0)$.

Use the points $(0, 4)$ and $(-1, 0)$ to draw the line.

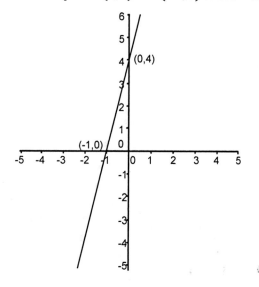

Second graph the equation $y = x - 1$.

Make a table of values or use intercepts to graph the line.

Let us use intercepts.

$x = 0 \rightarrow \quad y = 0 - 1 \rightarrow \quad y = -1$; The y-intercept is $(0, -1)$.

$y = 0 \rightarrow \quad 0 = x - 1 \rightarrow \quad 1 = x$; The x-intercept is $(1, 0)$.

Use the points $(0, -1)$ and $(1, 0)$ to draw the second line.

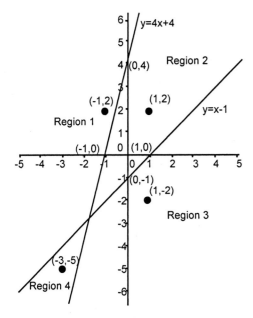

Also notice that the two lines form four regions. Choose arbitrary points in these regions to determine the solutions or points that make **both inequalities true**. The points arbitrarily chosen are:

Region 1 $\rightarrow (-1, 2)$

Region 2 $\rightarrow (1, 2)$

Region 3 $\rightarrow (1, -2)$

Region 4 $\rightarrow (-3, -5)$

Test all the points in both equations. The point that makes both inequalities true is the point that will be in the shaded area.

➢ Testing the point $(-1, 2)$ in region 1, we have:

$y > 4x + 4$ and $y \leq x - 1$

$2 > 4(-1) + 4$ and $2 \leq -1 - 1$

$2 > -4 + 4$ and $2 \leq -2$

$2 > 0$ true and $2 \leq -2$ false $\rightarrow$ the point $(-1, 2)$ in region 1 **does not work** for both inequalities.

➢ Testing the point $(1, 2)$ in region 2, we have:

$y > 4x + 4$ and $y \leq x - 1$

$2 > 4(1) + 4$ and $2 \leq 1 - 1$

$2 > 4 + 4$ and $2 \leq 0$

$2 > 8$ false and $2 \leq 0$ false → The point $(1, 2)$ in region 2 **does not work** for both inequalities.

➢ Testing the point $(1, -2)$ in region 3, we have:

$y > 4x + 4$ and $y \leq x - 1$

$-2 > 4(1) + 4$ and $-2 \leq 1 - 1$

$-2 > 8$ and $-2 \leq 0$

$-2 > 8$ false and $-2 \leq 0$ true → The point $(1, -2)$ in region 3 **does not work** for both inequalities.

➢ Testing the last point $(-3, -5)$ in region 4, we have:

$y > 4x + 4$ and $y \leq x - 1$

$-5 > 4(-3) + 4$ and $-5 \leq -3 - 1$

$-5 > -12 + 4$ and $-5 \leq -4$

$-5 > -8$ true and $-5 \leq -4$ true → The point $(-3, -5)$ in region 4 **works for both inequalities.**

The graph below is shaded in region 4 where the point $(-3, -5)$ is located. One inequality is drawn using the dotted line ($>$) and the other one is drawn using a solid line ($\leq$). Please note that if the point $(-3, -5)$ is tested first and it works, there would be no need to test all the other points. Also, shade only the **darkest area**. It is not necessary to shade the other areas as shown in this graph.

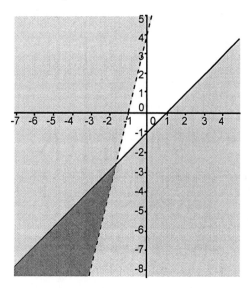

<u>Additional</u> Problems for <u>selected</u> sections

These can also be used during lectures.

❖ **Section A.1 – Solve absolute value equations.**

1) $2|x - 5| = 6$

2) $3|x + 4| - 5 = 10$

3) $|8y| + 2 = 34$

4) $\left|\frac{x}{3} - 5\right| + 4 = 2$

5) $\left|\frac{2-7c}{3}\right| = 1$

6) $11 + 7|5p| = 14$

<u>Answers</u>: 1) $2, 8$ 2) $-9, 1$ 3) $-4, 4$ 4) No solutions   5) $-\frac{1}{7}, \frac{5}{7}$ 6) $-\frac{3}{35}, \frac{3}{35}$

❖ **Section A.2 – Solve compound inequalities (Set builder and interval notations).**

1) $4x - 2 \geq 6$ and $-2x > 4$

2) $-4 \leq 5(x + 1) \leq 7$

3) $\frac{1}{7} < \frac{5-2x}{14} < \frac{3}{7}$

4) $2 - 5x > -8$ and $-x + 10 > 11$

5) $6x \geq 18$ or $-\frac{2}{11}x - 7 > 5$

6) $-x \leq -4$ or $2x + 1 \leq 15$

<u>Answers</u>: 1) No solutions or $\emptyset$ 2) $\left\{x \left| -\frac{9}{5} \leq x \leq \frac{2}{5}\right.\right\}; \left[-\frac{9}{5}, \frac{2}{5}\right]$ 3) $\left\{x \left| -\frac{1}{2} < x < \frac{3}{2}\right.\right\}; \left(-\frac{1}{2}, \frac{3}{2}\right)$

4) $\{x \mid x < -1\}; (-\infty, -1)$ 5) $\{x \mid x < -66$ or $x \geq 3 \}; (-\infty, -66) \cup [3, \infty)$

6) $\{x \mid x \leq 7$ or $x \geq 4 \};$ or all real numbers or $(-\infty, \infty)$

❖ **Section 1.1 – Solve applications of intercepts.**

1) Valencia College found that it takes 1 hour to provide support (advising, testing, registration, ATLAS, etc.) to returning students and 2 hours to new students. A total of 600 hours is available. This scenario can be expressed as the linear equation, $x + 2y = 600$, where x is the number of returning students and y is the number of new students.
   a. How many new students can be helped if there are no returning students?
   b. How many returning students can be helped if there are no new students?

2) The Osceola Math Depot found that it takes approximately 0.5 hour or 30 minutes to provide help to algebra students and approximately 2 hours to Calculus students. A total of 72 hours is available in a given week. This scenario can be expressed as the linear equation, $0.5x + 2y = 72$, where x is the number of algebra students who need help and y is the number calculus students who need help.
   a. How many algebra students can be helped if no calculus students come for help?
   b. How many calculus students can be helped if no algebra students come for help?

<u>Answers</u>: 1) a. 300   b. 600   2) a. 144   b. 36

❖ Section 1.2 – Slope of a line

Find the slope given points.

1) $\left(0,\frac{1}{2}\right)$ and $\left(\frac{3}{5},1\right)$ 2) $\left(-\frac{2}{7},-\frac{1}{9}\right)$ and $\left(-\frac{3}{7},-\frac{2}{9}\right)$ 3) $\left(-7,\frac{2}{3}\right)$ and $\left(-7,\frac{4}{5}\right)$ 4)$\left(-10,\frac{2}{11}\right)$ and $\left(5,\frac{2}{11}\right)$

Find the slope and y-intercept of the lines.

5) $y = 7 - 2x$ 6) $-2x - 5y = 11$ 7) $10y - 6 = 2x$ 8) $y = -12x$ 9) $x - 9 = 0$

10) $y + 15 = 0$ 11) $y = 0.25x + 5.3$ 12) $x = 7 - 9y$

Determine if the lines are parallel, perpendicular, or neither.

13) $4x - 2y = -20$ and $4x + 8y = 4$ 14) $y = -\frac{3}{10}x + 1$ and $y = \frac{3}{10}x - 2$

15) $-12x + 6y = 15$ and $6x - 3y = 21$ 16) $y = \frac{1}{2}x - 11$ and $y = \frac{1}{2}x + 15$

Answers: 1) $\frac{5}{6}$ 2) $\frac{7}{9}$ 3) undefined 4) 0   5)-2; $(0,7)$ 6) $-\frac{2}{5}$; $\left(0,-\frac{11}{5}\right)$ 7)$\frac{1}{5}$; $\left(0,\frac{3}{5}\right)$

8) -12; $(0,0)$ 9) undefined; no y-intercept   10) 0; $(0,-15)$ 11) 0.25; $(0,5.3)$

12) $-\frac{1}{9}$; $\left(0,\frac{7}{9}\right)$ 13) perpendicular   14) neither   15) parallel   16) parallel

❖ Section 1.3 – Equations of a line

Find an equation of the line given the following information.

1) $m = -11$; $(-1,-7)$ 2) $m = \frac{2}{5}$; $\left(-\frac{3}{5},-\frac{1}{2}\right)$ 3)$\left(\frac{1}{7},\frac{2}{9}\right)$ and $\left(-\frac{2}{7},\frac{5}{9}\right)$ 4) $(2.1,7.4)$and $(0.15,9.1)$

5) $(1,9)$; parallel to $y = 3 - 7x$ 6) $(3,11)$; parallel to $1 + 9y = -2x$

7) $(-1,3)$; perpendicular to $y = 3 - 7x$ 8) $(-3,-1)$; perpendicular to $1 + 9y = -2x$

9) The revenue for one of the Valencia bookstores went up at a constant rate of $2100 every year since 2008. The bookstore's revenue in 2008 was $37,500. Find an equation that expresses the revenue since 2008.

10) At Valencia College, due to budget constraints, the funds available to student development activities have decreased at a constant rate of $3500 per year since 2010. The funds that were available in 2010 were $250,000. Find an equation that expresses the decrease.

11) The Dual enrollment program at Valencia has experienced a constant growth since 2001. In 2003, 910 students dual enrolled at Valencia. In 2007, 926 students dual enrolled at Valencia. Use this information to write an equation of the line that models this situation. Use x as the number of years after 2001 and y as the number of students.

12) Use the equation in number 11) to predict the number of students who will dual enroll in 2015.

Answers: 1) $y = -11x - 18$ 2) $y = \frac{2}{5}x - \frac{13}{50}$ 3) $y = -\frac{7}{9}x + \frac{1}{3}$ 4) $y = -0.87x + 9.23$

5) $y = -7x + 16$ 6) $y = -\frac{2}{9}x + \frac{35}{3}$ 7) $y = \frac{1}{7}x + \frac{22}{7}$ 8) $y = \frac{9}{2}x + \frac{25}{2}$

9) $y = 2100x + 37500$ 10) $y = -3500x + 250000$ 11) $y = 4x + 902$ 12) 958

❖ **Section 1.4 – Find domain and range (Set builder and interval notations).**

1) $X^2 + Y^2 = 16$

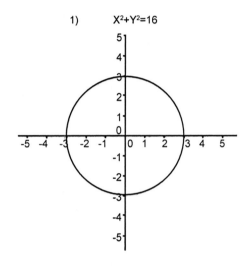

2) $y = x^2$

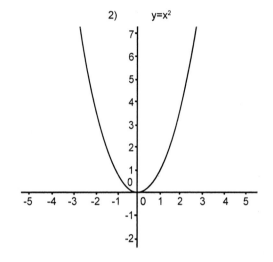

3) $(x-2)^2 + (y+3)^2 = 4$

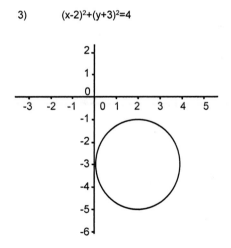

4) $(y+3)^2 = (x+4)$

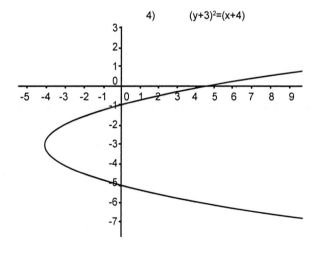

5) X²-Y²=9

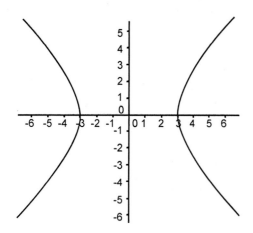

6) $y=\sqrt[3]{x}$

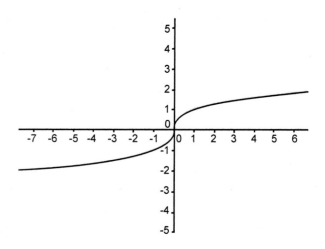

Answers: 1) Dom: $\{x \mid -4 \leq x \leq 4\}$; $[-4,4\,]$ Ran: $\{y \mid -4 \leq y \leq 4\}$; $[-4,4\,]$

2) Dom: all real numbers or $(-\infty, \infty)$ Ran: $\{y \mid y \geq 0\}$; $[0, \infty)$

3) Dom: $\{x \mid 0 \leq x \leq 4\}$; $[0,4\,]$ Ran: $\{y \mid -5 \leq y \leq -1\}$; $[-5,-1\,]$

4) Dom: $\{x \mid x \geq -4\}$; $[-4, \infty)$ Ran: all real numbers or $(-\infty, \infty)$

5) Dom: $\{x \mid x \leq -3 \text{ or } x \geq 3\}$; $(-\infty, -3] \cup [3, \infty)$ Ran: all real numbers or $(-\infty, \infty)$

6) Dom and Ran: all real numbers or $(-\infty, \infty)$

❖ **Section 1.5 – Functions**

Is this relation a function?

1) $\left\{\left(0, \frac{1}{4}\right), (-1, 2), \left(3, \frac{2}{5}\right), (4, -1)\right\}$ 2) $\{(0.1,\ 4), (1,\ 2.7), (3,\ 5.1), (4, -1), (3, -6.2)\}$

Evaluate the functions.

The function $P(x) = -\frac{1}{3}x^2 + 2x - 5$ is given. Find:

3) $P(-3)$ 4) $P\left(\frac{1}{2}\right)$ 5) $P\left(-\frac{2}{3}\right)$ 6) $P(7.5)$

Find the domain of these functions.

7) $S(x) = 5x^3 + 4x + 10$ 8) $M(x) = \dfrac{2x}{2x-3}$ 9) $F(x) = \dfrac{x^2-4x}{x-\frac{1}{2}}$ 10) $S(x) = \dfrac{3-x}{4x+\frac{5}{4}}$

11) $D(x) = \sqrt{3x - 7}$ 12) $H(x) = \sqrt{1 - \frac{x}{2}}$ 13) $I(x) = \sqrt[3]{x + 6}$ 14) $K(x) = \sqrt[4]{2x - 1}$

15) $I(x) = \sqrt[3]{x - 2.5}$

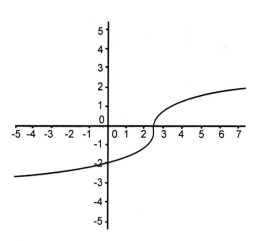

16) $I(x) = \sqrt[4]{x - 2}$

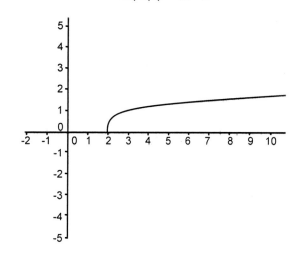

17) $I(x) = \sqrt[5]{2x - 3}$

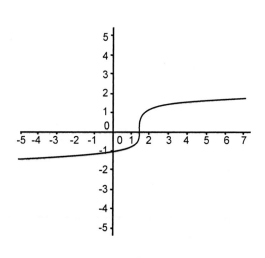

18) $R(x) = \dfrac{3}{x - 1}$

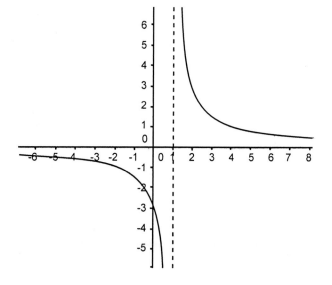

Solve (Use 3.1416 for π).

19) The side of a square is 10 centimeters. Find the area of a square. $[A(s) = s^2]$

20) The radius of a sphere is 7 inches. Find the volume of the sphere. $[V(r) = \frac{4}{3} \pi r^3]$

21) The radius of a circle is 13 inches. Find the circumference of the circle. $[C(r) = 2\pi r]$

22) The side of an equilateral triangle is 9 inches. Find the area of the triangle. $[A(s) = \frac{\sqrt{3}}{4} s^2]$

Answers: 1) yes   2) No   3) -14 4) $-\frac{49}{12}$ 5) $-\frac{175}{27}$ 6) -8.75

7) All real numbers or $(-\infty, \infty)$ 8) $\{x \mid x \neq \frac{3}{2}\}$; $\left(-\infty, \frac{3}{2}\right) \cup \left(\frac{3}{2}, \infty\right)$ 9) $\{x \mid x \neq \frac{1}{2}\}$; $\left(-\infty, \frac{1}{2}\right) \cup \left(\frac{1}{2}, \infty\right)$

10) $\{x \mid x \neq -\frac{5}{16}\}$; $\left(-\infty, -\frac{5}{16}\right) \cup \left(-\frac{5}{16}, \infty\right)$ 11) $\left\{x \mid x \geq \frac{7}{3}\right\}$; $\left[\frac{7}{3}, \infty\right)$ 12) $\{x \mid x \leq 2\}$; $(-\infty, 2]$

13) All real numbers or $(-\infty, \infty)$ 14) $\left\{x \mid x \geq \frac{1}{2}\right\}$; $\left[\frac{1}{2}, \infty\right)$ 15) All real numbers or $(-\infty, \infty)$

16) $\{x \mid x \geq 2\}$; $[\, 2, \infty)$ 17) All real numbers or $(-\infty, \infty)$ 18) $\{x \mid x \neq 1\}$; $(-\infty, 1) \cup (1, \infty)$

19) 100 cm²   20) $1{,}436.76$ in³   21) 81.68 in   22) 35.07 in²

❖ **Sections 2.3 and 2.4 – Systems of linear equations/Elimination or Substitution**

Solve using either method.

1) $\frac{2}{3}x + 2y = \frac{8}{3}$

   $-\frac{1}{2}x - y = -\frac{1}{2}$

2) $\frac{9}{4}x - \frac{3}{2}y = -\frac{3}{2}$

   $3x + 3y = -\frac{9}{2}$

3) $\frac{3}{2}x + 5y = 22$

   $\frac{1}{8}x - \frac{3}{2}y = -2$

4) $2x + \frac{3}{4}y = -\frac{9}{2}$

   $\frac{3}{2}x - \frac{3}{4}y = -6$

5) $12x - 21y = 21$

   $36x - 63y = 70$

6) $2x = 6y + 4$

   $10x - 30y = 20$

7) $10x - 5y = -5$

   $5y = -10x$

8) $1.4x - 0.4y = -3.2$

   $0.4x - 2y = -2.8$

9) Two travelers start at the same location and travel in opposite directions. One travels 8 miles faster than the other. In 8 hours, they are 560 miles away from each other. Find the speed of each traveler.

10) Jason has a 10% saline solution and a 60% saline solution. How many liters of each solution must be mixed to make 50 liters of a 40% saline solution?

Answers: 1) $(-5, 3)$ 2) $\left(-1, -\frac{1}{2}\right)$ 3) $(8, 2)$ 4) $(-3, 2)$ 5) No solutions   6) Infinite   7) $\left(-\frac{1}{4}, \frac{1}{2}\right)$ 8) $(-2, 1)$

9) 31 mph and 39 mph   10) 10% solution – 20 liters; 60% solution – 30 liters

❖ **Sections 3.4 – Long Division of Polynomials**

1) $(5y^2 - 9) \div (y - 2)$

2) $(8x^3 + 8x^2 - 2x + 1) \div \left(x + \frac{3}{2}\right)$

3) $(x^3 + 64) \div (x + 4)$

4) $\left(y^4 - \frac{2}{5}y^3 + y\right) \div (y - 1)$

Answers: 1) $5y + 10 + \frac{11}{y-2}$ 2) $8x^2 - 4x + 4 - \frac{5}{x+\frac{3}{2}}$ 3) $x^2 - 4x + 16$

4) $y^3 + \frac{3}{5}y^2 + \frac{3}{5}y + \frac{8}{5} + \frac{\frac{8}{5}}{y-1}$ or $y^3 + \frac{3}{5}y^2 + \frac{3}{5}y + \frac{8}{5} + \frac{8}{5y-5}$

❖ **Sections 3.6 – Factoring the sum and difference of cubes**

1) $\frac{8}{27}x^3 - \frac{1}{64}$

2) $\frac{1}{125}y^3 + \frac{1}{216}$

3) $128x^3 - 250y^3$

4) $108a^3 + 32b^3$

5) $x^3y^3 - 125$

6) $m^3n^3 + 27$

<u>Answers:</u> 1) $\left(\frac{2}{3}x - \frac{1}{4}\right)\left(\frac{4}{9}x^2 + \frac{1}{6}x + \frac{1}{16}\right)$

2) $\left(\frac{1}{5}y + \frac{1}{6}\right)\left(\frac{1}{25}y^2 - \frac{1}{30}y + \frac{1}{36}\right)$

3) $2(4x - 5y)(16x^2 + 20xy + 25y^2)$

4) $4(3a + 2b)(9a^2 - 6ab + 4b^2)$

5) $(xy - 5)(x^2y^2 + 5xy + 25)$

6) $(mn + 3)(m^2n^2 - 3mn + 9)$

❖ **Sections 4.1 – Simplifying, finding domains, and evaluating**

Simplify the rational expressions.

1) $\frac{5x^2-5x}{5x^3-10x^2+5x}$

2) $\frac{7a^2-7a}{7a^2-21a+14}$

3) $\frac{P}{7p^2+9P}$

4) $\frac{x^2-x-42}{7x-x^2}$

5) $\frac{2x^2-9x+10}{6x^3-15x^2+2x-5}$

Find the domain of the rational functions.

6) $f(x) = \frac{3+5x}{6x^2+x-15}$

7) $g(x) = \frac{1+x}{16x^2-49}$

8) $h(x) = \frac{x^2-x-9}{x^3+6x^2-7x}$

9) $p(x) = \frac{x+11}{9x}$

10) $f(x) = \frac{7x}{6x^2+\frac{2}{3}}$

11) $f(a) = \frac{3}{a^2-9a}$

Evaluate the rational function.

12) $f(x) = \frac{x+3}{x^2-4}$; find   a) $f(-2)$ b) $f(-3)$ c) $f\left(\frac{1}{2}\right)$ d) $f\left(-\frac{3}{4}\right)$

<u>Answers:</u> 1) $\frac{1}{x-1}$ 2) $\frac{a}{a-2}$ 3) $\frac{1}{7p+9}$ 4) $-\frac{x+6}{x}$ 5) $\frac{x-2}{3x^2+1}$

6) $\left\{x \mid x \neq -\frac{5}{3},\ x \neq \frac{3}{2}\right\}; \left(-\infty, -\frac{5}{3}\right) \cup \left(-\frac{5}{3}, \frac{3}{2}\right) \cup \left(\frac{3}{2}, \infty\right)$

7) $\left\{x \mid x \neq -\frac{7}{4},\ x \neq \frac{7}{4}\right\}; \left(-\infty, -\frac{7}{4}\right) \cup \left(-\frac{7}{4}, \frac{7}{4}\right) \cup \left(\frac{7}{4}, \infty\right)$

8) $\{x \mid x \neq -7,\ x \neq 0, x \neq 1\}; (-\infty, -7) \cup (-7,0) \cup (0, 1) \cup (1, \infty)$ 9) $\{x \mid x \neq 0\}; (-\infty, 0) \cup (0, \infty)$

10) All real numbers or $(-\infty, \infty)$ 11) $\{a \mid a \neq 0,\ a \neq 9\}; (-\infty, 0) \cup (0,9) \cup (9, \infty)$

12) a) undefined   b) 0   c) $-\frac{14}{15}$ d) $-\frac{36}{55}$

❖ **Sections 4.2 – Multiplying and dividing rational expressions**

1) $\dfrac{24x^4}{20y^3} \cdot \dfrac{25y^6}{8x}$

2) $\dfrac{x-y}{x^2} \cdot \dfrac{x^2y}{x^2-3xy+2y^2}$

3) $\dfrac{2x^2-6x}{x^2-9} \cdot \dfrac{7xy^2-7y^3}{2x^2-2xy}$

4) $\dfrac{c^2-10c+25}{3c^2-14c-5} \cdot \dfrac{c+7}{c^3-5c^2}$

5) $\dfrac{4a^2+36}{4a^2-20} \div \dfrac{a^4-81}{a^4-8a^2+15}$

6) $\dfrac{2b^2-11b+5}{4b^2+12b-7} \div \dfrac{2b^2-50}{2b^2+9b+7}$

7) $\dfrac{2y+10}{3x+21} \div \dfrac{xy+5x-3y-15}{x^2-6x+9}$

8) $\dfrac{2x^2-32}{x^3+4x^2+x+4} \div \dfrac{7x^2}{x^3+x}$

Answers: 1) $\dfrac{15x^3y^3}{4}$　2) $\dfrac{y}{x-2y}$　3) $\dfrac{7y^2}{x+3}$　4) $\dfrac{c+7}{c^2(3c+1)}$　5) $\dfrac{a^2-3}{a^2-9}$　6) $\dfrac{b+1}{2(b+5)}$

7) $\dfrac{2(x-3)}{3(x+7)}$　8) $\dfrac{2(x-4)}{7x}$

❖ **Sections 4.3 – Adding and subtracting rational expressions**

1) $\dfrac{x^3}{x-4} - \dfrac{64}{x-4}$

2) $\dfrac{y^3}{y+5} + \dfrac{125}{y+5}$

3) $\dfrac{x-y}{xy^2} + \dfrac{x+y}{y^2x}$

4) $\dfrac{2}{y^2-49} - \dfrac{1}{y^2+7y}$

5) $\dfrac{4x}{3x-6} + \dfrac{8}{6-3x}$

6) $\dfrac{2x}{x^3-8} - \dfrac{5}{x^2+2x+4}$

Answers:　1) $x^2+4x+16$　2) $y^2-5y+25$　3) $\dfrac{2}{y^2}$　4) $\dfrac{1}{y(y-7)}$　5) $\dfrac{4}{3}$　6) $\dfrac{-3x+10}{x^3-8}$

❖ **Sections 4.4 – Simplifying complex fractions**

1) $\dfrac{\frac{x^2}{16}-1}{-\frac{1}{2}+\frac{x}{8}}$

2) $\dfrac{\frac{7}{2x}}{\frac{5}{3x}-6}$

3) $\dfrac{\frac{1}{x+2}+\frac{4}{x-2}}{\frac{7}{x-2}-\frac{2}{x+2}}$

4) $\dfrac{\frac{4}{x-3}+\frac{2}{x+3}}{\frac{5}{x^2-9}}$

5) $\dfrac{x^{-2}}{x^{-3}-y^{-3}}$

Answers: 1) $\dfrac{x+4}{2}$　2) $\dfrac{21}{10-36x}$　3) $\dfrac{5x+6}{5x+18}$　4) $\dfrac{6(x+1)}{5}$　5) $\dfrac{xy^3}{y^3-x^3}$

❖ **Sections 4.5 – Solving rational and literal equations**

1) $\dfrac{1}{x+1} + \dfrac{2}{x} = \dfrac{4}{7x}$

2) $\dfrac{3}{x+2} + \dfrac{1}{x-2} = \dfrac{7x-19}{x^2-4}$

3) $\dfrac{4x-5}{7x} + \dfrac{9}{7x} = 4$

4) $\dfrac{6}{y} = 5 - y$

5) $\dfrac{b-3}{2b+1} = \dfrac{2}{3}$

6) $\dfrac{10}{a^2-25} = \dfrac{1}{a-5}$

7) The difference of one fourth a number and its reciprocal is equal to 32 divided by the number. Find the number(s).

8) Marie and Levi can each rake the leaves in the yard in 2 hours and 3 hours respectively. Owen can do the same job alone in 4 hours. If Marie, Levi, and Owen work together, how long will it take them to rake the leaves in the yard?

9) It would take Rebecca 9 months to write a calculus book working alone. It was estimated that it could take just 6 months to write the calculus book if she and Taj work together. How long would it take Taj to write the book alone?

Solve for the indicated variable.

10) $\dfrac{a}{b+c} = \dfrac{d}{c}$; c

11) $a = \dfrac{by}{dy+c}$; y

12) $\dfrac{1}{x} + \dfrac{1}{y} = \dfrac{1}{z}$; z

13) $S = \dfrac{n(a+L)}{2}$; L

Answers: 1) $-\dfrac{10}{17}$ 2) 5 3) $\dfrac{1}{6}$ 4) 2, 3 5) -11 6) No solutions

7) $-12, 12$ 8) $\dfrac{12}{13}$ Hr. 9) 18 months 10) $c = \dfrac{db}{a-d}$ 11) $y = \dfrac{-ac}{ad-b}$

12) $z = \dfrac{xy}{y+x}$ 13) $L = \dfrac{2S-na}{n}$

❖ **Sections 5.5 – Solving radical equations**

1) $\sqrt[3]{x} + 2 = 13$

2) $\sqrt[4]{x+7} - 2 = 3$

3) $\sqrt[5]{y-2} + 7 = -1$

4) $\sqrt[6]{2x+1} = 2$

5) $\sqrt{2+x} - 1 = 2x$

6) $\sqrt{3x+1} - 1 = \sqrt{2x-1}$

7) $\sqrt{5y-1} - \sqrt{y} + 1 = 2$

8) The formula for instantaneous velocity after elapsed time t is given by $v = \sqrt{2gd}$, where d is the distance in meters that the object falls and g is gravity in meter per second squared, approximately 9.8 m/sec^2. Find the distance that an object has fallen if its velocity is 75 m/sec.

9) After an accident, the speed that a car was traveling can be estimated using the measurement of its skid marks. The formula $S=\sqrt{21d}$ can be used, where S is the speed, in miles per hour, and d is the length of the skid marks, in feet.
If a car is traveling at 45 mph and the driver brakes suddenly, how much distance is required for the car to come to a complete stop?

10) The formula for the period of a pendulum is given by $T = 2\pi \sqrt{\dfrac{l}{g}}$, where T is the period in seconds (time required for one complete vibration), l is the length of the pendulum in feet, and g is the acceleration due to gravity (32 ft./s^2). Find the length of the pendulum which period is 5 seconds. Use 3.1416 for π.

Answers: 1) 1331   2) 618   3)−32,766   4)$\dfrac{63}{2}$ 5)$\dfrac{1}{4}$ 6) 1, 5   7) 1   8) 287 meters   9) 96 ft.   10) 20.26 ft.

❖ **Sections 6.1 – Solve by factoring**

1) $x(x - 2) = 24$

2) $2x(2x - 1) = 6x - 3$

3) $3x(x + 2) - 2 = x$

2) $y^2 - 16y + 20 = -43$

5) $a^2 + 9a - 33 = a$

6) $b^2 - 8b = -15$

7) $2x^2 - \dfrac{5}{2} = \dfrac{19}{2}x$

8) $x^3 = 49x$

9) In a right triangle, one leg is 7 feet longer that the other leg. The hypotenuse is 17 ft. Find the length of the two legs of the triangle.

10) In a right triangle, one leg is 7 feet longer that the other leg. The hypotenuse is 13 ft. Find the length of the two legs of the triangle.

11) A purse accidently dropped from a hot air balloon can be modeled by this function, $h(t) = -16t^2 + 400$, where h(t) is the height above the ground and t is the time in seconds. When will the purse hit the ground?

Answers: 1) $-4, 6$ 2)$\dfrac{1}{2}, \dfrac{3}{2}$ 3)$-2, \dfrac{1}{3}$ 4) 7, 9   5) $-11, 3$ 6) 3, 5   7) $-\dfrac{1}{4}, 5$ 8)$-7, 0, 7$

9) 8 ft. and 15 ft.   10) 5 ft. and 12 ft.   11) 5 sec

❖ **Sections 6.2 – Solve by using the square root method**

1) $(y + 2)^2 - 49 = 0$

2) $(a - 5)^2 - 24 = 0$

3) $b^2 + 11 = 44$

2) $3x^2 - 12 = -x^2 + 36$

5) $12(1 - x^2) + 11(x^2 - 2) = -54$

6) $4(x^2 - 5)^2 + 2 = 326$

7) $6(2x - 3)^2 - 7 = 11$

8) $(3y + 2)^2 + 64 = 0$

9) The formula for calculating the total amount of money, A, with annual compound interest is given by $A = P(1 + r)^t$, where P is the initial amount, r is the interest rate, and t is the time. Nadeen invested $2500 and this amount grew to $2702 in 2 years. Find the interest rate.

10) The formula for calculating the total amount of money, A, with annual compound interest is given by $A = P(1 + r)^t$, where P is the initial amount, r is the interest rate, and t is the time. Mandy Liz invested $900 and this amount grew to $1150 in 2 years. Find the interest rate.

11) Ignoring air resistance, the distance in feet of a falling object is given by this function, $d(t) = 16t^2$, where t is the time in seconds. The Orange County Courthouse in Orlando is 416 feet. How long would it take an object to fall if dropped from the top of the courthouse?

Answers: 1) $-9, 5$ 2) $5 \pm 2\sqrt{6}$ 3) $\pm\sqrt{33}$ 4) $\pm 2\sqrt{3}$ 5) $\pm 2\sqrt{11}$ 6) $\pm\sqrt{14}, \pm 2i$ 7) $\dfrac{3\pm\sqrt{3}}{2}$

8) $\dfrac{-2\pm 8i}{3}$ 9) 3.9 or 4%   10) 13%   11) 5.1 sec

❖ **Sections 6.4 – Solve using the quadratic formula**

1) $x(x - 2) = 25$ 2) $2x(2x - 1) = x + 3$ 3) $3x(x + 2) + 2 = 2x$

2) $t^2 - 7t + 20 = 12$ 5) $5a^2 + 9a - 3 = a$ 6) $b^2 - 8b = -10$

7) $\dfrac{1}{12}x^2 - \dfrac{1}{3}x = -\dfrac{1}{6}$

Answers: 1) $1 \pm \sqrt{26}$ 2) $\dfrac{3\pm\sqrt{57}}{8}$ 3) $\dfrac{-2\pm i\sqrt{2}}{3}$ 4) $\dfrac{7\pm\sqrt{17}}{2}$ 5) $\dfrac{-4\pm\sqrt{31}}{5}$ 6) $4 \pm \sqrt{6}$

7) $2 \pm \sqrt{2}$

❖ **Sections 6.6 – Find the vertex of quadratic functions**

1) $g(x) = 6x^2 + 6x + 2$ 2) $g(x) = 3x^2 + x + 5$ 3) $h(x) = \dfrac{1}{5}x^2 - 3x - 1$ 4) $h(x) = -\dfrac{2}{3}x^2 + \dfrac{1}{2}x - 3$

5) *Solve.* If a ball is kicked from the ground with an initial velocity of 24 ft. per second, then the function can be expressed as $f(x) = -16t^2 + 24t$. Find the maximum height of the ball.

6) *Solve.* Valencia students are trying to raise funds for a trip by selling hot dogs, and the function that can be used is $A(x) = -x^2 + 80x$. Find the maximum amount of money raised.

Answers: 1) $\left(-\dfrac{1}{2}, \dfrac{1}{2}\right)$ 2) $\left(-\dfrac{1}{6}, \dfrac{59}{12}\right)$ 3) $\left(\dfrac{15}{2}, -\dfrac{49}{4}\right)$ 4) $\left(\dfrac{3}{8}, -\dfrac{93}{32}\right)$ 5) 9 ft.   6) $1,600

✎ **Answers to "Your turn" Exercises**

Section A.1 - Absolute value equations

1. $-8, 8$
2. $-12, -2$
3. $-\frac{7}{3}, \frac{11}{3}$
4. $-48, 48$
5. $-1, \frac{17}{3}$
6. No solutions
7. $-\frac{4}{5}, -\frac{14}{3}$
8. $-14, 18$

Section A.2 – Compound inequalities

1. $\{x \mid x < -2\}$; $(-\infty, -2)$
2. No solutions
3. $\{x \mid -8 < x < 5\}$; $(-8, 5)$
4. $\{x \mid -7 \le x \le -2\}$; $[-7, -2]$
5. $\left\{x \mid 2 < x < \frac{19}{2}\right\}$; $(2, \frac{19}{2})$
6. $\left\{x \mid x \le \frac{13}{2} \text{ or } x \ge 9 \right\}$; $\left(-\infty, \frac{13}{2}\right] \cup [9, \infty)$
7. $\{x \mid x < 8 \text{ or } x > -5 \}$; $(-\infty, \infty)$
8. $\{x \mid x < 5\}$; $(-\infty, 5)$

Section 1.1 – Linear equations

1. No
2. No
3. Yes

4. Intercepts: $(1/4, 0)$ and $(0, -1/2)$

5. Intercepts: $(6,0)$ and $(0, -2)$

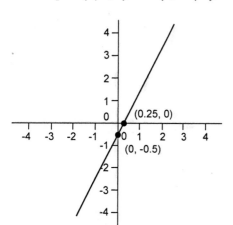

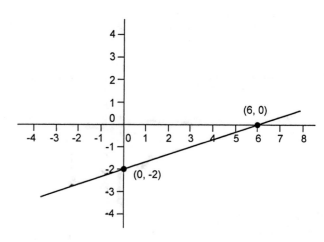

6. Intercepts: (4, 0) and (0, −2)

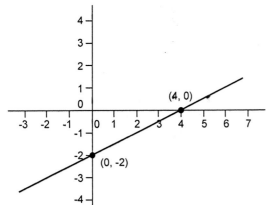

7. Intercepts: (−5, 0) and (0, 2)

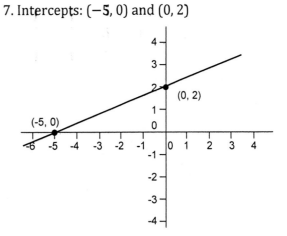

8. $x = -2$

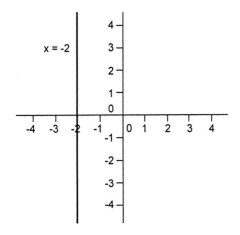

9. $x = 4$

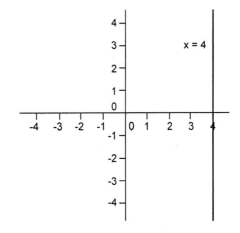

10. $y = 2$

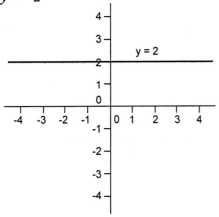

11. $y = -3$

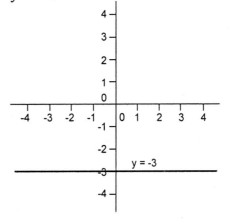

12. a) $840.00 b) 21months

Section 1.2 – Slope of a line

1. 2
2. −2
3. 0
4. Undefined

5. 3/5
6. −9/11
7. −6.4 or − 6 (rounding). The number of students decreased by 6 every year.
8. The slope is 892. The enrollment number increased by 892 students every year.
9. Slope −2 and y-intercept (0, 2/5)
10. Slope −5/4 and y-intercept (0, 9/4)
11. Slope 3 and y-intercept (0, 18)
12. Perpendicular
13. Parallel
14. Neither
15. Slope is ½ and y-intercept is (0, −4)   16. Slope is −4/5 and y-intercept is (0,2)

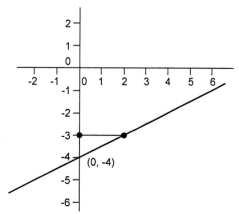 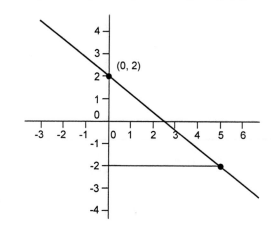

Section 1.3 – Equations of the line

1. $y = 3x + 8$
2. $y = -\frac{1}{7}x + \frac{51}{7}$
3. $y = x + 1$
4. $y = \frac{4}{5}x + \frac{19}{5}$
5. $y = -6x - 8$
6. $y = -\frac{1}{2}x + \frac{7}{2}$
7. $y = 3x + 14$
8. $y = \frac{1}{3}x + \frac{5}{3}$
9. $a)\ y = 11.5x + 15\ \ b)\$61$
10. $y = 13x + 29;\ \$68$
11. $y = 48x - 96,159$
12. $y = 16x - 31,070$

Section 1.4 – Relations

1. Domain: $\{-7, -6, -5, 4, 0, 2\}$; Range: $\{9, 6, 5, -6, -9, -1\}$
2. Domain: $\{0, 1, 2, 3, 4\}$; Range: $\{2, 5, 6, -2, -7, -3\}$
3. Domain: $\{0, 1, 2, 5\}$; Range $\{5, 6, 3, -1\}$
4. Domain: $\{x | x \geq -2\}$ or $[-2, \infty)$; Range $\{y | y \leq 0\}$ or $(-\infty, 0]$.
5. Domain: All real numbers or $(-\infty, \infty)$; Range $\{y | y \geq 2\}$ or $[2, \infty)$.

Section 1.5 – Functions

1. Yes
2. No
3. No
4. Yes
5. No
6. $w(-1) = 10$
7. $P(-2) = -38$
8. $R\left(\frac{1}{3}\right) = -3$
9. $R\left(\frac{1}{7}\right) = -\frac{25}{7}$
10. a) $f(0) = 4$; b) $f(-3) = 4$; c) $x = -2,\ x = 3, x = -4$
11. a) \$57; b) 11 hours
12. a) \$4700; b) 17.5 months
13. The domain consists of all real numbers or $(-\infty, \infty)$
14. The domain is $\{x | x \neq -3\}$ or $(-\infty, -3) \cup (-3, \infty)$.
15. The domain is $\{x | x \neq 2\}$ or $(-\infty, 2) \cup (2, \infty)$.
16. The domain is $\{x | x \neq -5/4\}$ or $(-\infty, -5/4) \cup (-5/4, \infty)$.
17. The domain is $\{x | x \geq -5\}$ or $[-5, \infty)$.
18. The domain is $\{x | x \leq 1\}$ or $(-\infty, 1]$.

Section 2.1 - Systems of linear equations - introduction

1. No
2. Yes
3. Yes
4. No
5. No

Section 2.2 - Systems of linear equations - graphing

System 1 – No solutions

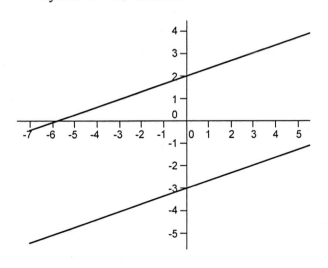

System 2 – Infinitely many solutions

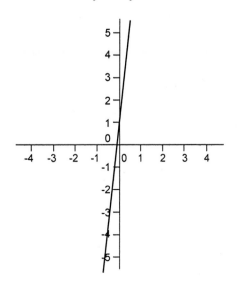

System 3 – One solution (4.5, 2.5)

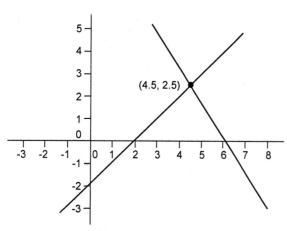

System 4 – One solution (4, 3)

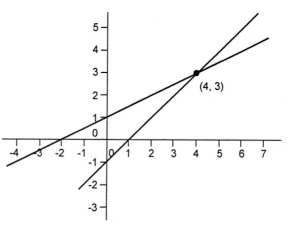

Section 2.3 - Systems of linear equations - elimination

System 1 → no solutions

System 2 → infinitely many solutions

System 3 → one solution (1, 5)

System 4 → one solution (8, -3)

System 5 → children $3; adult $5

System 6 → test 4 points each; quiz 1 point each

System 7→ faculty 120; students 130

Section 2.4 - Systems of linear equations - substitution

System 1 → no solutions

System 2 → infinitely many solutions

System 3 → one solution (1, 5)

Systems 4 → infinitely many solutions

Systems 5 → the numbers are 10 and 15.

System 6 → Pascal 6 hours; Regis 3 hours

System 7→ $4000 at 3%; $9000 at 6%

Section 2.5 – Graphing linear inequalities and systems of linear inequalities

1.

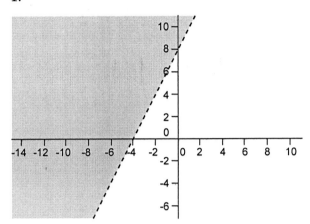

2.

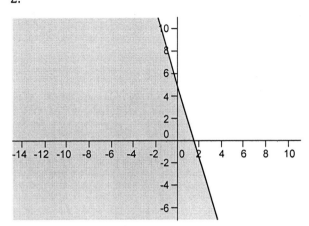

3.

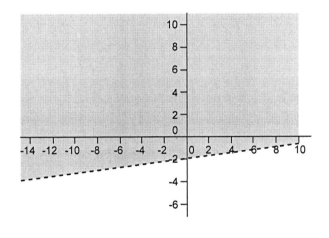

System 1

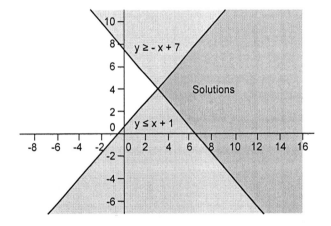

System 2

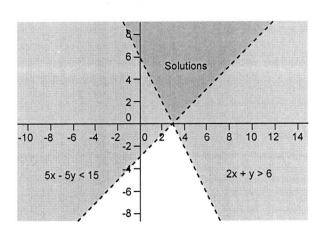

System 3

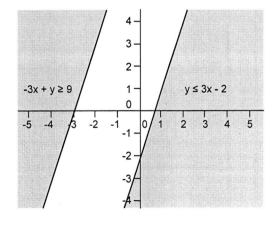

Section 3.1 – Introducing polynomials

1. $-10x^3, 18x^2, \frac{1}{5}x \, ; \frac{1}{5}$
2. 3 ; trinomial
3. $\frac{1}{3}x^5 + 3x^4 + 2x^3 + 6x^2 - 5x - \frac{5}{7}; \frac{1}{3}x^5; \frac{1}{3}$
4. $-20x^3, 8x^2, -7x \, ; -7$
5. 4; binomial
6. $17x^5 + 13x^4 + 4x^3 + 3x^2 - 11x - 1; 17x^5; 17$

Section 3.2 – Adding and subtracting polynomials

1. $-3y^7 + 2y^5 + 2y^3 + 20y^2$
2. $-3m^5n^4 - 9m^3n^2 + 5mn$
3. $\frac{5}{6}s^3 + \frac{1}{3}s^2 + 4s + \frac{32}{5}$
4. $-5y^7 + 7y^5 - 11y^3 - 8y^2$
5. $15m^5n^4 + 11m^3n^2 + 12mn$
6. $-\frac{1}{2}s^3 - s^2 - s + \frac{28}{5}$
7. $P = 8x + 14$
8. $3x + 6$

Section 3.3 – Multiplying polynomials

1. $18x^2yz^4$
2. $-3m^5n^{11}p^6$
3. $a^{12}b^{22}c^{16} - \frac{1}{2}a^{11}b^{28}c^{20}$
4. $-2x^3y + 9x^2y^4 - \frac{7}{4}x^2y$
5. $x^2 - 2x - 15$
6. $20x^2 - 39x + 7$
7. $y^2 - 36$
8. $9x^2 - 16y^2$
9. $4x^2 - 20x + 25$
10. $\frac{1}{16}x^2 + x + 4$
11. $A = 3x^2 + 17x + 10$
12. $A = x^2 + 4x + 4$
13. $x = 12$ cm

Section 3.4 – Dividing polynomials (Long division)

1. $x - 7$
2. $x - 2$
3. $x + 9$
4. $2x + 1$

5. $4x - 14 + \dfrac{84}{x+5}$

6. $3x - 17 + \dfrac{47x+123}{x^2+4x+7}$

7. $x^2 + 6x + 38 + \dfrac{196x-179}{x^2-6x+5}$

Section 3.5 – Factoring polynomials

1. $6(x^2 + 5y^4 - 8z^5)$
2. $3m^2n^3(4m^4n^4 + 5)$
3. $7abc^3(2b^2c^2 - 4ac + 9)$
4. $(x + 4)(x^2 + 5)$
5. $(3y - 4)(y^3 + y - 5)$
6. $\left(k + \frac{1}{2}\right)\left(\frac{1}{7}k - 9\right)$
7. $(x + 5)(x - 3)$
8. $(2y - 3)(y - 7)$
9. $(k + 3)(7k - 6)$
10. $(x + 4)(x + 2)$
11. $(x + 7)(x - 8)$
12. $(x - 2)(x - 7)$
13. $(x + 6)^2$
14. $(x + 3)(2x + 1)$
15. $(3x - 7)(2x - 1)$
16. $(4x + 3)(x - 1)$
17. $15(y + 1)^2$
18. Prime or not factorable
19. $(7x - 4y)(7x + 4y)$
20. $\left(\frac{2}{5}x - \frac{4}{11}\right)\left(\frac{2}{5}x + \frac{4}{11}\right)$
21. $2xy(4x - 3y)(4x + 3y)$
22. Prime or not factorable
23. $(6 + y)(6 - y)$
24. $(c + d)(c - d)$
25. $(3x - 2)^2$
26. $(2x + 5)^2$
27. $(6x + 7)^2$
28. $(x - 2)(x + 2)(x^2 + 4)$

Section 3.6 – Factoring polynomials: sum and difference of cubes

1. $(y - 6)(y^2 + 6y + 36)$
2. $(3y - 2)(9y^2 + 6y + 4)$
3. $2(x - 3)(x^2 + 3x + 9)$
4. $(x + 4)(x^2 - 4x + 16)$

5. $(x + 1)(x^2 - x + 1)$

6. $(4y + 5)(16y^2 - 20y + 25)$

Section 4.1 – Simplifying, finding domains, & evaluating

1. $\dfrac{1}{2}$

2. $\dfrac{x-2}{x+5}$

3. $\dfrac{2x+3}{x+6}$

4. $\dfrac{-1}{9+x}$

5. $-\dfrac{3}{5}$

6. $\dfrac{2}{3y+15}$

7. $\{x \mid x \neq 5\}$ or $(-\infty, 5) \cup (5, \infty)$

8. $\{x \mid x \neq -3, x \neq 2\}$ or $(-\infty, -3) \cup (-3, 2) \cup (2, \infty)$.

9. $\{x \mid x \neq -4, x \neq 5\}$ or $(-\infty, -4) \cup (-4, 5) \cup (5, \infty)$.

10. All real numbers or $(-\infty, \infty)$

11. $0, \dfrac{2}{3}, \dfrac{5}{3}$; $-\dfrac{12}{5}$

Section 4.2 – Multiplying & dividing rational expressions

1. $\dfrac{4}{x(x-6)}$

2. $\dfrac{6k}{k-5}$

3. $\dfrac{4}{5}$

4. $x + 2$

5. $\dfrac{5}{9}$

6. $\dfrac{-1}{(x-12)(x+5)}$

7. $\dfrac{x^2+x+1}{2x}$

8. $\dfrac{x^2-2x+4}{x^2-2x}$

Section 4.3 – Adding & subtracting rational expressions

1. $\dfrac{8}{x}$

2. $\dfrac{2x-15}{x^3}$

3. $(x+7)$

4. $\dfrac{x+5}{x}$

5. $\dfrac{x-12}{x^2}$

6. $\dfrac{12x+50}{30x}$ or $\dfrac{6x+25}{15x}$

7. $\dfrac{-4}{x(x-4)}$

8. $\dfrac{x^2+5x+5}{(x+4)(x+3)}$

9. $\dfrac{x^2-15x+19}{(x-3)(x+4)(x-1)}$

Section 4.4 – Simplifying complex fractions

1. $\dfrac{2m+4}{5m-12}$

2. $n-7$

3. $n-9$

4. $\dfrac{x^2(2x+3)}{x+9}$ or $\dfrac{2x^3+3x^2}{x+9}$

5. x^2

6. $\dfrac{3x+2}{2}$ or $\dfrac{2+3x}{2}$

Section 4.5 – Solving rational equations & literal equations

1. $x=\dfrac{20}{3}$
2. $x=-6,7$
3. $k=\dfrac{21}{10}$
4. $x=-\dfrac{1}{7}$
5. The denominator of the equation is zero when $s=1$. Correct answer is **no solution**.
6. $t=\dfrac{6}{5}$ or 1.2 hours
7. $t=\dfrac{15}{8}$ or approximately 2 months
8. Luciano 51 mph and Toby 66 mph
9. $B=AC-C$ or $B=C(A-1)$

10. $\dfrac{M}{MP-1} = N$

11. $M = \dfrac{FR}{V^2}$

12. $\dfrac{ER^2}{K} = Q$

Section 5.1 – Simplifying radical expressions & evaluating radical functions

1. 5
2. 9
3. 12
4. $\dfrac{2}{9}$
5. -5
6. Not a real number
7. 2
8. 2
9. -3
10. Not a real number
11. $\dfrac{1}{3}$
12. $\dfrac{2}{3}$
13. 5
14. 5
15. 12
16. 6
17. -4
18. 2
19. 2
20. y
21. $x + 2$
22. $5\sqrt{2}$
23. $-2\sqrt[3]{2}$
24. $7xy^3\sqrt{xy}$
25. $10lm^2n^3\sqrt{n}$
26. $-3a^2b^4c^5\sqrt[3]{c}$
27. $3vwy^2\sqrt[4]{u^3w^2y}$
28. Approximately 27 feet
29. Approximately 22 feet
30. a) 5.916   b) 2.571   c) 2.088   d) 1.615
31. Answers for f: 2, $\sqrt{5}$, 3; answers for g: $\sqrt[3]{5}$, -2, 3 ; answers for h: 2, $\sqrt[4]{2}$, 1

Section 5.2 – Working with scientific, standard notation, and rational exponents

1. a) 1.34×10^{-5} b) 7.5×10^{7}
2. a) 2,350,000 b) 0.000439
3. a) 5

   b) ¾

   c) 3

   d) −4

   e) Not a real number

   f) 2

   g) 1/7

   h) 1/2

   i) −1/2
4. a) 9

   b) 4/9

   c) 216

   d) 8

   e) 81

   f) 1/16

   g) −1/27

   h) Not a real number

5. a) 1.72×10^{4}

   b) 3.69×10^{3}

   c) 2.135×10^{2}

   d) 7.04×10^{-4}

   e) 1.23×10^{13}

   f) 1.25×10^{2}

   g) 6.21×10^{-1}

   h) 4×10^{-10}

6.

   a) $x^{1/20}$

   b) x^{2}

   c) $x^{13/21}$

   d) $\dfrac{1}{x^{2}}$

   e) $\sqrt[30]{x^{11}}$

   f) $\sqrt[6]{11}$

   g) $x + x^{13/6}$

   h) $x^{\frac{4}{10}} - 7x^{\frac{11}{10}} + 4x^{31/10}$

   i) $x + 12x^{\frac{1}{2}} + 35$

7. a) $x^{\frac{-1}{2}}(4 + 11x^4)$

    b) $x^{\frac{1}{3}}(1 - x^2)$

    c) $x^{\frac{-3}{5}}(\frac{1}{4} - 9x^{\frac{6}{5}} + \frac{1}{10}x^{\frac{2}{5}})$

Section 5.3 – Adding & subtracting radical expressions

1. $6\sqrt{5}$
2. $\frac{101}{60}\sqrt{a}$
3. $9\sqrt{2}$
4. $9x^4\sqrt{3x}$
5. $(4p + 3)\sqrt[3]{p}$
6. $5\sqrt{3} + \sqrt{15}$
7. $-\frac{4}{3}\sqrt[4]{3}$
8. $-4\sqrt{np}$
9. $-11k^2\sqrt{K}$
10. $-12y\sqrt[6]{y}$

Section 5.4 - Multiplying and dividing radical expressions.

1. $2\sqrt{3}$
2. $2\sqrt{5}$
3. $2\sqrt[3]{10}$
4. $10y\sqrt{2}$
5. $\frac{3}{5}\sqrt{2}$
6. $2 + 2\sqrt{2}$
7. $8 - 2\sqrt{10}$
8. $2\sqrt{10} + \sqrt{30}$
9. $2\sqrt{2} + \sqrt{6} + 2\sqrt{5} + \sqrt{15}$
10. $\sqrt{3} - 4 + 3\sqrt{2} - 4\sqrt{6}$
11. -44
12. $10 + 8\sqrt{3}$
13. $x - 17$
14. $x - 23$
15. $\frac{4}{5}$
16. $\sqrt{3}$
17. 6
18. $\frac{2}{5}$

19. $3x^2\sqrt{x}$

20. $\dfrac{9\sqrt{11}}{11}$

21. $\dfrac{7+\sqrt{3}}{46}$

Section 5.5 – Solving radical equations

1. 36
2. -125
3. 256
4. 9
5. 18
6. No solution
7. -134
8. $-\dfrac{2}{3}$
9. -13
10. No solution
11. $x = 7$
12. $x = 19$
13. 13
14. 552.9 cm^3
15. 113.1 cm^3

Section 5.6 - Working with complex numbers

1. $7i$
2. $2i\sqrt{10}$
3. $i\sqrt{11}$
4. $-\sqrt{21}$
5. $-2\sqrt{6}$
6. -3
7. $\dfrac{4}{5}$
8. $\dfrac{\sqrt{10}}{2\sqrt{5}}$ or $\dfrac{\sqrt{2}}{2}$
9. $\dfrac{\sqrt{5}}{\sqrt{6}}$ or $\dfrac{\sqrt{30}}{6}$
10. a) i
    b) 1
    c) -1
    d) 1
11.
    a) $11 - 9i$
    b) $11 + 3i$
    c) $-1 + 8i$

d) $-2 + 7i$

e) -30 or $-30 + 0i$

f) 35 or $35 + 0i$

g) $6 + 2i$

h) 101 or $101 + 0i$

i) $11 - 2i$

j) $143 + 24i$

k) $\dfrac{11 - 8i}{5}$ or $\dfrac{11}{5} - \dfrac{8}{5}i$

l) $\dfrac{-19 + 17i}{50}$ or $-\dfrac{19}{50} + \dfrac{17}{50}i$

m) $\dfrac{-11i}{5}$ or $0 - \dfrac{11}{5}i$

Section 6.1 - Solving quadratic equations - factoring

1. $x = -4, -2$

2. $x = -7, 8$

3. $x = 2, 7$

4. $x = -3, -\dfrac{1}{2}$

5. $x = \dfrac{1}{2}, \dfrac{7}{3}$

6. $-\dfrac{3}{4}, 1$

7. $y = -1$

8. $x = -\dfrac{2}{5}, \dfrac{2}{5}$

9. $x = -\dfrac{10}{11}, \dfrac{10}{11}$

10. $x = 0, 13$

11. $w = 5$ inches; $l = 12$ inches

12. 4 seconds

13. $0 and $300

Section 6.2 - Solving quadratic equations - square root method

1. $x = \pm 11$

2. $x = \pm 2\sqrt{5}$

3. $x = \pm \sqrt{7}$

4. $x = -8, 6$

5. $x = \dfrac{9}{4}, \dfrac{15}{4}$

6. $x = -1, 5$

7. $x = \dfrac{-4 \pm \sqrt{6}}{3}$

8. $x = \pm 8i$

9. $x = \pm 7$

10. $r = 12.62$ cm

11. $x = -7 \pm \sqrt{5}$

12. $t = 1.85$ seconds

Section 6.3 - Solving quadratic equations - complete the square

1. $x = 2 \pm \sqrt{7}$
2. $x = 5 \pm \sqrt{26}$
3. $x = \frac{5}{2} \pm \frac{\sqrt{13}}{2}$ or $x = \frac{(5 \pm \sqrt{13})}{2}$
4. $x = 1 \pm \sqrt{7}$
5. $-\frac{1}{2}, 2$

Section 6.4 - Solving quadratic equations - quadratic formula

1. $x = \frac{1 \pm \sqrt{13}}{2}$
2. $x = \frac{-4 \pm \sqrt{6}}{2}$
3. $x = \frac{1}{2}, 2$
4. $x = \frac{-1 \pm i\sqrt{55}}{14}$
5. $x = \frac{1 \pm \sqrt{5}}{2}$
6. $x = -1, \frac{-2}{3}$
7. $w = 3.7$ inches; $l = 6.7$ inches
8. $t = 3.45$ seconds
9. $t = 0.72$ and 2.78 seconds
10. 32 and 34

Section 6.5 – Graphing quadratic functions

1

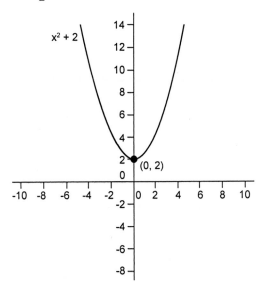

2

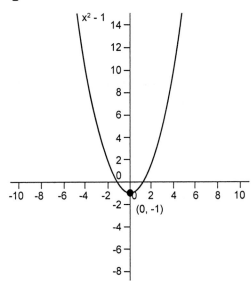

3

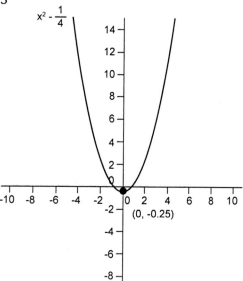

$x^2 - \dfrac{1}{4}$

(0, -0.25)

4

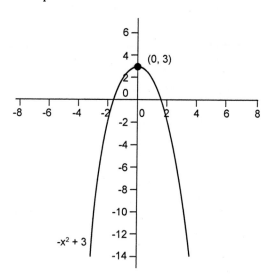

(0, 3)

$-x^2 + 3$

5

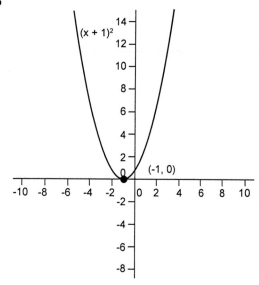

$(x + 1)^2$

(-1, 0)

6

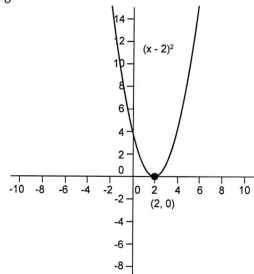

$(x - 2)^2$

(2, 0)

7

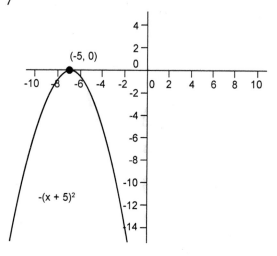

(-5, 0)

$-(x + 5)^2$

8

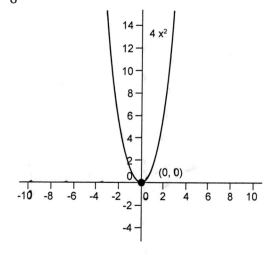

$4x^2$

(0, 0)

9

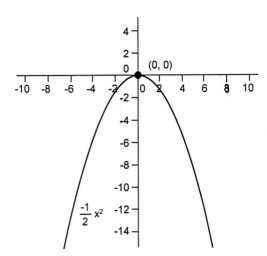

10

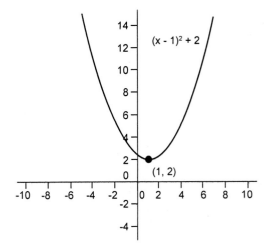

11

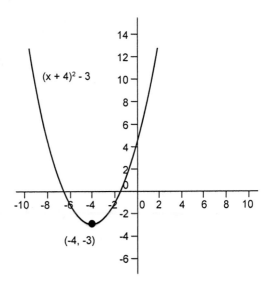

12

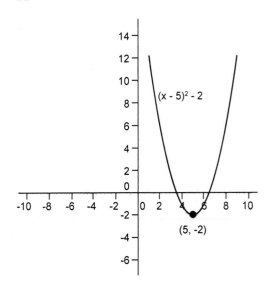

13

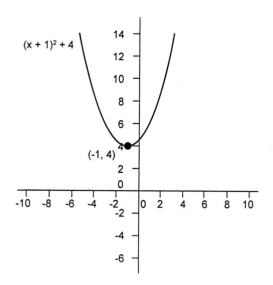

14

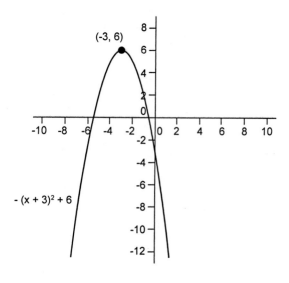

Section 6.6 – Finding the vertex of quadratic functions

1. $(8, 3)$

2. $(-1, -4)$

3. $(5, -2)$

4. $(-7, 4)$

5. $(4, -12)$

6. $(3, -13)$

7. $(-3, -23)$

8. $(2, 7)$

9. $(-2, -3)$

10. 158 feet

11. $306,250

12. 18 cookies; $176

INDEX

T

V

X

Y